# Interactive
## Mathematics Program®

INTEGRATED HIGH SCHOOL MATHEMATICS

# Orchard Hideout

FIRST EDITION AUTHORS:
Dan Fendel, Diane Resek, Lynne Alper, and Sherry Fraser

CONTRIBUTORS TO THE SECOND EDITION:
Sherry Fraser, IMP for the 21st Century
Jean Klanica, IMP for the 21st Century
Brian Lawler, California State University San Marcos
Eric Robinson, Ithaca College, NY
Lew Romagnano, Metropolitan State College of Denver, CO
Rick Marks, Sonoma State University, CA
Dan Brutlag, Meaningful Mathematics
Alan Olds, Colorado Writing Project
Mike Bryant, Santa Maria High School, CA
Jeri P. Philbrick, Oxnard High School, CA
Lori Green, Lincoln High School, CA
Matt Bremer, Berkeley High School, CA
Margaret DeArmond, Kern High School District, CA

Key Curriculum Press

Second Edition    I M P

This material is based upon work supported by the National Science Foundation under award numbers ESI-9255262, ESI-0137805, and ESI-0627821. Any opinions, findings, and conclusions or recommendations expressed in this publication are those of the authors and do not necessarily reflect the views of the National Science Foundation.

Key Curriculum Press
1150 65th Street
Emeryville, California 94608
email: editorial@keypress.com
www.keypress.com
10 9 8 7 6 5 4 3 2 1    14 13 12 11
ISBN 978-1-60440-047-2
Printed in the United States of America

Project Editors
Mali Apple, Josephine Noah, Sharon Taylor

Project Administrators
Emily Reed, Juliana Tringali

Professional Reviewers
Rick Marks, Sonoma State University, CA
D. Michael Bryant, Santa Maria High School, CA, retired

Accuracy Checker
Carrie Gongaware

First Edition Teacher Reviewers
Daniel R. Bennett, Moloka'i High School, HI
Maureen Burkhart, Northridge Academy High School, CA
Dwight Fuller, Ponderosa High School, CA
Daniel S. Johnson, Silver Creek High School, CA
Brian Lawler, California State University San Marcos, CA
Brent McClain, Vernonia School District, OR
Susan Miller, St. Francis of Assisi Parish School, PA
Amy C. Roszak, Cottage Grove High School, OR
Carmen C. Rubino, Silver Creek High School, CA
Barbara Schallau, East Side Union High School District, CA
Kathleen H. Spivack, Wilbur Cross High School, CT
Wendy Tokumine, Farrington High School, HI

First Edition Multicultural Reviewers
Genevieve Lau, Ph.D., Skyline College, CA
Arthur Ramirez, Ph.D., Sonoma State University, CA
Marilyn Strutchens, Ph.D., Auburn University, AL

Copyeditor
Brandy Vickers

Interior Designer
Marilyn Perry

Production Editor
Andrew Jones

Production Director
Christine Osborne

Editorial Production Supervisor
Kristin Ferraioli

Compositor
Lapiz Digital Services, Kristin Ferraioli

Art Editor/Photo Researcher
Maya Melenchuk

Technical Artists
Lapiz Digital Services, Laurel Technical Services, Maya Melenchuk

Illustrators
Taylor Bruce, Deborah Drummond, Tom Fowler, Briana Miller, Evangelia Philippidis, Sara Swan, Diane Varner, Martha Weston, April Goodman Willy

Cover Designer
Jeff Williams

Printer
Lightning Source, Inc.

Mathematics Product Manager
Elizabeth DeCarli

Executive Editor
Josephine Noah

Publisher
Steven Rasmussen

# CONTENTS

# Orchard Hideout

Circles and Coordinate Geometry

# Orchard Hideout—Circles and Coordinate Geometry

# Orchards and Mini-Orchards

Have you ever stared into an orchard as you passed it on the road? From one perspective, you see that the trees are planted in straight rows. But as you move to a different spot, all you see is a mass of trees.

The main characters of this unit, Madie and Clyde, have planted an orchard. They want to know how long it will take the trees to grow before they can no longer see from the center of the orchard to the outside world.

You will begin the unit by looking at their overall problem, and then you'll examine some simpler cases. You will also explore some activities related to the first POW of the unit.

*Annie Tam and Ryan Tran work to develop the mathematical definition for a circle.*

# Orchard Hideout

Madie and Clyde want some peace and quiet. Most of all, they want privacy. So they leave the city and buy a piece of land in the countryside.

Their lot is in the shape of a circle. They decide to plant an orchard on the lot in nice neat rows.

Here's how they set up their orchard.

They plant their first row of trees along an east-west line through the center of the circle. They space the trees equally, except they leave out the tree that would have been at the circle's exact center. There are 50 trees to the east of the center and 50 to the west. The trees at the ends of this east-west row are exactly on the boundary of the property.

Madie and Clyde then plant a north-south line of trees through the center, using the same spacing as before and omitting the tree at the center. Again, there are 50 trees to the north of the center and 50 to the south. And the trees at the ends of this north-south row are exactly on the boundary of the property.

They use each tree in the north-south row as the center of an east-west row, filling in the orchard with rows of trees. They use the same distance between trees in every row.

Madie and Clyde realize that as the trees grow, their trunks will become so big that it will eventually be impossible to see out from the center of the orchard. The center of the orchard would then be like a hideout.

*continued* ▶

Here is the main question of this unit.

*How soon after Madie and Clyde plant their orchard will the center of the lot become a true "orchard hideout"?*

1. Study the problem. Make a model of the situation, perhaps using a smaller orchard as an example.

2. Make a list of questions you need to ask to understand this problem better. Try to answer some of your questions.

Adapted from "The Orchard Problem" in *Mathematical Gems I,* by Ross Honsberger (Washington, DC: Mathematical Association of America, 1973).

# *A Geometric Summary*

As you can see, the central problem of this unit involves circles, straight lines, and all sorts of distances. To solve the problem, you will use many ideas from geometry, including trigonometry.

Here is a summary of basic definitions and essential principles from geometry that were presented in Years 1 and 2 of the Interactive Mathematics Program. Over the course of this unit, you will develop other key ideas.

## I. Polygon Angle Sums

The *angle sum property* for triangles is as follows:

*In any triangle, the sum of the measures of the angles is exactly 180°.*

One proof of this principle is based on properties of parallel lines, discussed in Part V.

In general, the sum of the measures of the angles of a polygon depends only on how many sides the polygon has. This sum is always a multiple of 180°.

## II. Similarity

Two polygons are considered similar if they have the same shape (though not necessarily the same size). Here is the formal definition.

**Definition** Two polygons are *similar* if their corresponding angles are equal and their corresponding sides are proportional in length.

Recall that when we say two angles are equal, we really mean their measures, or sizes, are equal.

Here are two basic principles for proving similarity of triangles.

*If the corresponding angles of two triangles are equal, then the triangles must be similar.*

*If the corresponding sides of two triangles are proportional in length, then the triangles must be similar.*

*continued ⟩*

The angle sum property for triangles leads to a simpler version of the *corresponding angles* principle.

*If two angles of one triangle are equal to two angles of another triangle, then the triangles must be similar.*

## III. Congruence

Congruence is a special case of similarity. Two polygons are considered congruent if they have the same shape and the same size. This means they are similar and the ratio of corresponding sides is 1:1. Here is the formal definition.

**Definition** Two polygons are *congruent* if their corresponding angles are equal and their corresponding sides are equal in length.

Here are two principles for proving congruence. The first applies to all polygons, and the second is specifically for triangles.

*If two polygons are similar and some pair of corresponding sides are equal in length, then the polygons must be congruent.*

*If two sides and the angle they form in one triangle are equal to the corresponding parts of another triangle, then the triangles must be congruent.*

## IV. Right Triangles

A right triangle is a triangle in which one of the angles is a right angle. The sides of a right triangle have special names.

**Definition** In a right triangle, the sides forming the right angle are the *legs*. The side opposite the right angle is the *hypotenuse*.

The relationship between the lengths of the sides of a right triangle is summed up in the **Pythagorean theorem.**

*In any right triangle, the sum of the squares of the lengths of the legs is equal to the square of the length of the hypotenuse.*

*continued* ▶

The trigonometric functions describe ratios within a right triangle. Here are the basic definitions.

> **Definition** If $\triangle ABC$ is a right triangle with a right angle at $C$, the trigonometric functions **sine, cosine,** and **tangent** for $\angle A$ are defined by these ratios.

$$\sin A = \frac{BC}{AB} \quad \cos A = \frac{AC}{AB} \quad \tan A = \frac{BC}{AC}$$

Each trigonometric function has an **inverse trigonometric function** that is used to determine an angle if the value of the trigonometric function is known. For example, if $x$ is a number between 0 and 1, the inverse sine of $x$ (written $\sin^{-1} x$) is the angle between 0° and 90° whose sine is $x$.

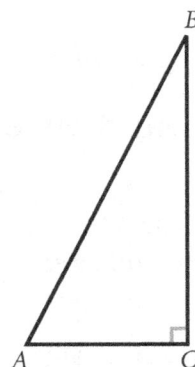

## V. Parallel Lines

Certain parts of a diagram like the one shown below have special names.

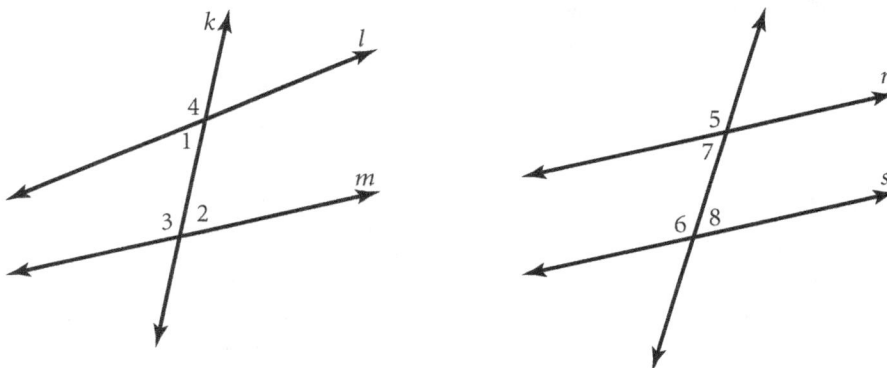

> **Definition** If two lines $l$ and $m$ are intersected by a third line $k$, the line $k$ is a *transversal*. A pair of angles like angles 1 and 2 are *alternate interior angles*. A pair of angles like angles 3 and 4 are corresponding angles.

Whether two lines intersected by a transversal are parallel is related to whether certain angles formed are equal.

For instance, in the diagram above, if lines $r$ and $s$ are parallel, then any pair of alternate interior angles (such as 7 and 8) must be equal, and

*continued* ◗

any pair of corresponding angles (such as 5 and 6) must be equal. Also, if any such pair of angles are known to be equal, then lines *r* and *s* must be parallel.

This gives us these principles.

> *If two parallel lines are intersected by a transversal, then any pair of alternate interior angles must be equal.*

> *If two parallel lines are intersected by a transversal, then any pair of corresponding angles must be equal.*

> *If a pair of alternate interior angles formed by a transversal intersecting two lines are equal, then the two lines must be parallel.*

> *If a pair of corresponding angles formed by a transversal intersecting two lines are equal, then the two lines must be parallel.*

*This Guatemalan weaver uses wooden sticks as transversals to cross the parallel lines of her thread.*

## VI. Perimeter, Area, Volume, and Surface Area

Here are several basic definitions regarding perimeter, area, volume, and surface area.

> **Definition** The *perimeter* of a polygon is the sum of the lengths of its sides.

> **Definition** The *area* of a plane figure is the number of square units it contains.

> **Definition** The *volume* of a solid figure is the number of cubic units it contains.

> **Definition** A *right prism* is a three-dimensional figure formed by moving a polygon through space in a direction perpendicular to the plane it lies in. The initial and final positions of the polygon are its *bases*. The other faces form the *lateral surface* of the prism.

*continued* ▶

The figures shown here are prisms. Two of them are right prisms. The first figure is called an *oblique prism* because it is created by a nonperpendicular movement of the base.

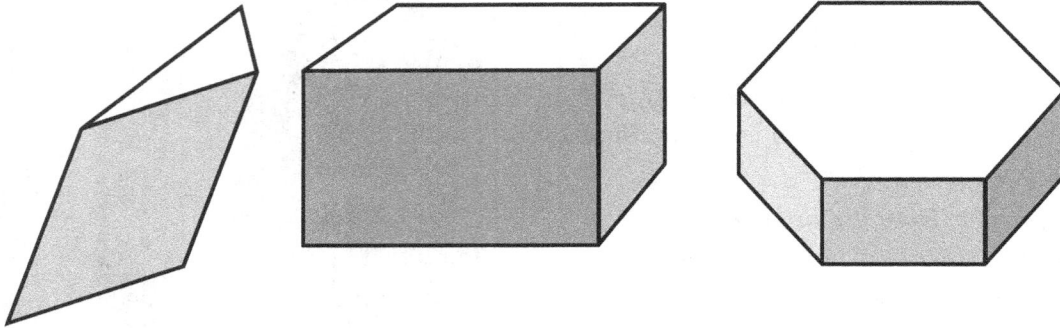

These principles describe basic formulas for finding area, volume, and surface area.

> *The area of a rectangle is equal to the product of its length and width.*
>
> *The area of a triangle is equal to half the product of its base and the corresponding altitude.*
>
> *The volume of a rectangular solid is equal to the product of its length, width, and height.*
>
> *The volume of a right prism is equal to the product of its height and the area of its base.*
>
> *The lateral surface area of a right prism is equal to the product of its height and the perimeter of its base.*

# Geometry and a Mini-Orchard

## Part I: Summarizing Geometry

Read *A Geometric Summary,* which reviews ideas about geometry and trigonometry that you will need during this unit. As you read, think about what each principle or definition means, and create diagrams to help you clarify the ideas.

## Part II: A Mini-Orchard

Imagine that Madie and Clyde's orchard is planted on a lot with a radius of 1 unit. At an early stage, this orchard might be represented by a diagram like this one.

Suppose you stand in the center of this orchard. If you look due east or north or west or south, a tree will block your line of sight. If you look in any other direction, you will be able to see out of the orchard through a gap between trees. As the trees grow, the gaps between them will narrow.

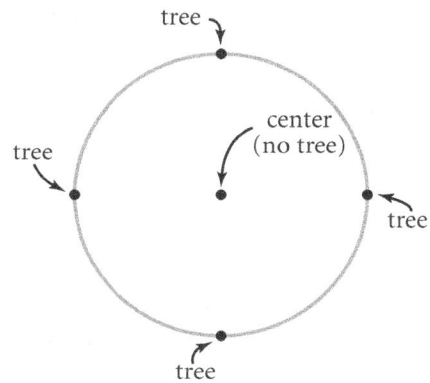

Now imagine that you remain at the center of the orchard, waiting patiently as the trees grow.

1. In what direction should you look to be able to see out of the orchard for as long as possible? That is, what line of sight will be the last to get blocked by the growing trees? If there is more than one answer, be as general as possible.

2. What is the minimum radius required for each tree trunk in order to make the center of the orchard into a true "orchard hideout"? In other words, how big must the trees become so that it is impossible for you to see out of the orchard from the center?

1. Two delicate flowers are planted in a garden. The gardener, Leslie, has a sprinkler that sprays water in a circle. The closer a flower is to the sprinkler, the more water it receives.

   To be sure her flowers each get the same amount of water, Leslie needs to place the sprinkler where it will be the same distance from each flower.

   What are her choices about where to put the sprinkler? Describe all the possibilities.

2. Leslie has another small garden where she wants to plant three flowers. She wants to know if it will be possible to place the sprinkler the same distance from all three.

   a. Determine which arrangements of the flowers (if any) will make this possible and which (if any) will make it impossible. As in Question 1, Leslie will be looking for a place to put the sprinkler after the flowers have been planted.

   b. For those arrangements for which it will be possible, describe how Leslie can find the correct location (or locations) for the sprinkler.

3. What about four flowers? Five flowers? Generalize as much as you can.

Your task is to explain as fully as possible, for various cases, where Leslie can place the sprinkler to give all the flowers the same amount of water. The activity *Only Two Flowers* will help get you started with Question 1 of this POW.

*continued*

○ *Write-up*

1. *Problem Statement:* State the problem in mathematical language without reference to the context. That is, describe the problem in geometric terms without talking about flowers or sprinklers.

2. *Process*

3. *Solution*

4. *Self-assessment*

Adapted from "Simple Math," a series of mathematics video-dramas produced by the Israeli Instructional Television Center, with academic adviser Nitsa Movshovitz-Hadar and producer Tamar Reiner.

Each Problem of the Week is unique, so the form of the write-up may vary from one POW to the next. Nevertheless, most of the categories you will use for your POW write-ups will be the same throughout the year. The list below summarizes the standard categories.

Some POW write-ups will use other categories or require more specific information within a particular category. If the write-up instructions for a given POW simply list a category by name, however, use these descriptions.

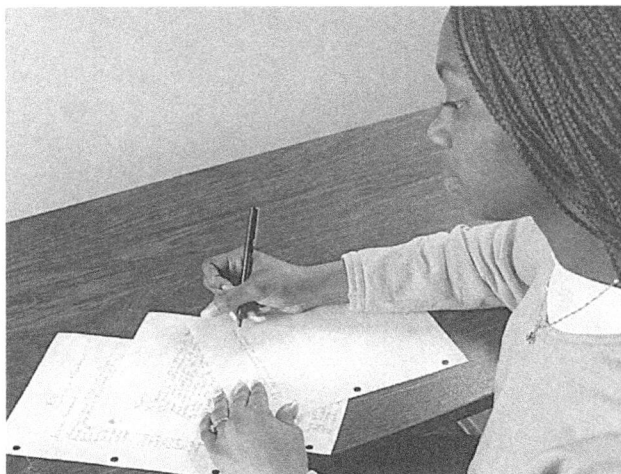

1. *Problem Statement:* State the problem clearly in your own words. Your problem statement should be clear enough that someone unfamiliar with the problem could understand what you are being asked to do.

2. *Process:* Describe what you did in attempting to solve the problem. Use your notes as a reminder. Include things that didn't work out or that seemed like a waste of time. Do this part of the write-up even if you didn't solve the problem.

   If you get assistance of any kind on the problem, tell what the assistance was and how it helped you.

3. *Solution:* State your solution as clearly as you can. Explain how you know your solution is correct and complete. If you obtained only a partial solution, give that. If you were able to generalize the problem, include your general results.

   Write your explanation in a way that will be convincing to someone else—even someone who initially disagrees with your answer.

4. *Self-assessment:* Tell what you learned from this problem. Be as specific as you can. Also assign yourself a grade for your work on this POW, and explain why you think you deserve that grade.

# Only Two Flowers

Your task is to answer the first question from the POW *Equally Wet*. That is, if two flowers are planted, what are the possible positions to place a sprinkler so that it is exactly the same distance from each flower?

Be sure your answer includes all possible solutions. Give your answer in two ways.

* With a diagram showing all the solutions
* With a description in words of the set of solutions

# A Perpendicularity Proof

Consider this principle.

*Every point on the perpendicular bisector of a line segment is equidistant from the two endpoints of the segment.*

Your task is to write a proof of this principle, based on the diagram shown here. In this diagram, $A$ and $B$ are any two points. Point $C$ is the midpoint of the line segment connecting $A$ and $B$. Line $l$ is the line through $C$ that is perpendicular to $\overline{AB}$ (so $l$ is the **perpendicular bisector** of $\overline{AB}$). Point $D$ is some point on $l$.

Based on this information, prove that $D$ is **equidistant** from $A$ and $B$. In other words, prove that lengths $AD$ and $BD$ are equal.

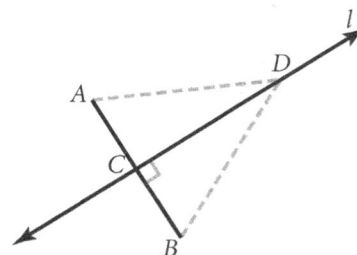

# From Two Flowers to Three

In *Only Two Flowers,* you determined where Leslie should place the sprinkler if she has two flowers to water. From that activity and its discussion, you should now have a simple description of Leslie's full set of options for locating the sprinkler for the case of two flowers.

Now consider a specific case using three flowers. The flowers are placed in a coordinate system at the points (4, 2), (14, 2), and (4, 8). Where can Leslie place the sprinkler so that it is equidistant from all three points? Explain in detail why the point you choose works.

# More Mini-Orchards

Madie and Clyde want to know how big the radius of each tree trunk needs to be for the orchard to become a hideout. They call this the "hideout tree radius."

Of course, this value depends on the radius of the orchard itself. In *Geometry and a Mini-Orchard* you examined the case of an orchard with a radius of 1 unit. But the orchard on a lot whose radius is 1 unit is pretty dull. Madie and Clyde are glad their lot is larger than that.

Suppose, though, that their orchard is just a *little* larger.

1. Sketch a mini-orchard for a lot with radius 2. Then answer these questions.
   a. How many trees are in this mini-orchard?
   b. Approximately how big must the radius of each tree trunk become to make this orchard a true hideout? In other words, approximately what is the hideout tree radius for an orchard of radius 2?

2. Sketch a mini-orchard for a lot with radius 3. Then answer these questions.
   a. How many trees are in this mini-orchard?
   b. Approximately what is the hideout tree radius for an orchard of radius 3?

3. Find the *exact* value of the hideout tree radius for orchards of radius 2 and radius 3.

# In, On, or Out?

It's convenient to use coordinates to describe situations like our orchard problem. Madie and Clyde plant a tree at every lattice point within the orchard, except at (0, 0). A **lattice point** is a point whose coordinates are both integers.

In using this system to describe the orchard, it's helpful to have a way to decide whether a given point is within the orchard. This is fairly easy for points on the axes. For other points, it can be more difficult.

For this activity, suppose the orchard has a radius of 10 (instead of 50). The center of the orchard is still at (0, 0).

Decide whether each point listed below is inside the boundary of the orchard, outside the boundary, or exactly on the boundary. In each case, explain how you decided. You may want to use diagrams to help you.

1. (11, 0)
2. (10, 0)
3. (10, 1)
4. (9, 3)
5. (9, 4)

6. (9, 5)
7. (9, 6)
8. (8, 5)
9. (8, 6)
10. (8, 7)

11. (7, 6)
12. (7, 7)
13. (7, 8)

## Coordinates and Distance

Madie and Clyde have many questions about the distance between trees and the distances from trees to various lines of sight.

By introducing coordinates into the orchard, you've made it possible to use the Pythagorean theorem to your advantage. In the upcoming activities, you will encounter some important formulas involving coordinates and distance.

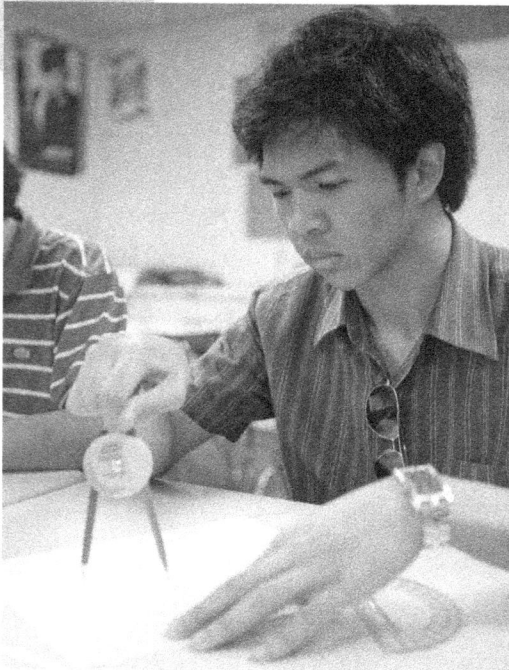

*Carlos Catly uses the coordinate system to model a mini-orchard.*

# Other Trees

1. Madie and Clyde buy another circular plot of land, smaller than the first, on which to plant an orchard. They have set up coordinates as before, with the center of the orchard at $(0, 0)$. They will plant trees at all points with integer coordinates that lie within the orchard, except at $(0, 0)$.

   In this orchard, the tree at $(5, 12)$ is on the boundary. What are the coordinates of the other trees that must also be on the boundary? Explain your answer.

2. Generalize Question 1 for a lot of any size. Suppose a circular orchard of any size is set up as usual with trees on the lattice points. If the point $(a, b)$ is on the lot's boundary, what other trees must also be on the boundary? Give as complete an answer as possible. You may want to examine other examples like that in Question 1.

# Sprinkler in the Orchard

Madie and Clyde are working in the scorching sun when they realize that their seedlings need water. So they haul out their trusty sprinkler and place it in the center of the orchard.

1. Suppose the sprinkler waters a circular area with a radius of 14 units. (Assume the trees do not block the water from one another.) State whether the tree at each of these locations will be watered, and explain your answers.

   a. $(-6, -13)$

   b. $(-9, 10)$

   c. $(-12, -8)$

2. Obviously, Madie and Clyde cannot reach all their trees when they place the sprinkler in the center of the orchard. After watering the trees near the center, they move the sprinkler to $(28, -19)$ to do some more watering. (Assume they temporarily uproot the tree at that location.) They also adjust the sprinkler so the water covers a circle of radius 18 units. Which of these trees will the water reach?

   a. $(16, -7)$

   b. $(38, -34)$

   c. $(20, -35)$

3. Assume the sprinkler continues to operate so that it reaches all trees within 18 units. In which locations can Madie and Clyde place the sprinkler to be able to water the entire orchard in as few waterings as possible?

   Work on this question from scratch, ignoring the waterings from $(0, 0)$ and $(28, -19)$ that are discussed in Questions 1 and 2.

# The Distance Formula

There is a well-known formula in mathematics, the **distance formula,** that is closely related to the Pythagorean theorem. This activity will help you develop that formula.

The distance formula is used to find the distance between two points in the coordinate plane, say, $(x_1, y_1)$ and $(x_2, y_2)$. The formula gives the distance between the two points in terms of the coordinates $x_1$, $x_2$, $y_1$, and $y_2$.

1.  a.  Find the distance between the points $(5, 3)$ and $(7, 6)$.

    b.  Describe in detail what you did to find your answer.

2.  a.  Find the distance between the points $(2, 5)$ and $(6, 3)$.

    b.  Describe in detail what you did to find your answer.

3.  Suppose $(x_1, y_1)$ and $(x_2, y_2)$ are the coordinates of two points. Generalize what you did in Questions 1 and 2 to create a formula or set of instructions that gives the distance between the two points.

4.  Does your generalization work if any of the coordinates $x_1$, $x_2$, $y_1$, and $y_2$ are negative numbers? Explain with examples.

# How Does Your Orchard Grow?

Madie and Clyde are trying to figure out how big the radius of the tree trunks must be for their orchard to become a hideout. They realize that once they figure this out, though, they still won't know how long it will take for the trees to grow to that size.

They return to the nursery where they bought the trees. The people at the nursery can't say when the orchard will become a hideout, but they are able to tell Madie and Clyde how fast the cross-sectional area of each tree trunk will increase.

When they get home, Madie and Clyde realize they don't know the current cross-sectional area of their trees. They go out to measure. Of course, they can't simply cut a cross section in a trunk and measure the area, because that would kill the tree. Instead, they measure the circumference of the trunk.

## Your Task

Imagine that Madie and Clyde know two things.

- The amount of increase each year of the cross-sectional area of the trees (This amount is the same every year.)
- The current circumference of the trees

Assume they also know the hideout tree radius. That is, they know how big the radius of each tree trunk needs to be for the orchard to become a hideout.

Develop a plan for how to use this information to figure out how long it will take the orchard to become a hideout. Identify any geometric questions you will need to answer to use your plan.

# A Snack in the Middle

Some of Madie and Clyde's trees already need pruning. Every afternoon they choose the two trees that need it the most, and each of them works on one of those two trees.

Pruning makes Madie and Clyde hungry. They decide to set up a snack table at the midpoint of the segment connecting the two trees they are working on. If there happens to be a tree at that exact point, they will set up the table right next to the tree.

1. Suppose Madie is working on the tree at (24, 6) and Clyde is working on the tree at (30, 14). What are the coordinates of the point where they should set up the snack table? Prove your choice is really equidistant from (24, 6) and (30, 14).

2. If they are working on the trees at (−3, 4) and (5, 12), where should they set up the table? Prove your choice is equidistant from (−3, 4) and (5, 12).

3. If they are working on the trees at (6, 2) and (11, −4), where should they set up the table? Prove your choice is equidistant from (6, 2) and (11, −4).

4. To save them time, make up a general formula for Madie and Clyde. Suppose the trees they are pruning are at $(x_1, y_1)$ and $(x_2, y_2)$. Find a formula for the midpoint between these two trees in terms of $x_1, y_1,$ $x_2,$ and $y_2$. This formula is called the **midpoint formula.**

# Equidistant Points and Lines

The POW *Equally Wet* involves the set of points equidistant from two or more given points. You're about to encounter some other problems that involve equal distances.

As you explore these activities, including your next POW, think about how these ideas might help Madie and Clyde with their orchard problem.

*Ryan Tran works on a proof that a given quadrilateral is a square.*

# Proving with Distance—Part I

1. This diagram shows a quadrilateral with vertices (5, 20), (9, 23), (12, 19), and (8, 16).

   Prove this figure is a square. (Don't simply show that the figure is a rhombus. Remember, a *rhombus* is a quadrilateral with four equal sides. A *square* is a rhombus with four right angles.)

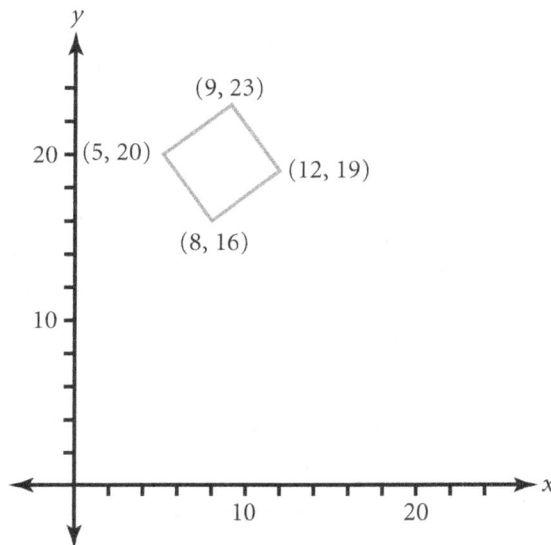

2. Choose a set of four points in the coordinate plane that form a quadrilateral. This can be any quadrilateral—it does not have to be a square. For simplicity, you may want to choose points in the first quadrant.

   a. Carefully plot the points and connect them to form a quadrilateral.

   b. Connect the midpoints of the sides of your quadrilateral to form a smaller quadrilateral.

3. Repeat Question 2 starting with a different set of four points.

4. Examine each of the smaller quadrilaterals you created.

   a. What general conjecture might you make about the lengths of the sides when you create a smaller quadrilateral this way?

   b. Find the coordinates of the midpoints in each of Questions 2 and 3. Verify your conjecture from Question 4a for each case using the distance formula.

# Down the Garden Path

Leslie, the gardener in the POW *Equally Wet,* has decided to plant only two flowers. She places them in her two favorite spots in the garden.

Now she has another idea. She wants to make a straight-line path through her garden, with one flower on each side of the path.

Leslie wants the two flowers to be the same distance from the path. That way, people walking along the path will see them both equally well (though not necessarily at the same time).

1. How can Leslie design a path that is equidistant from each flower? Write simple, step-by-step instructions for her.

2. Is your path the only one possible? Describe all possible straight-line paths that are equidistant from each flower.

3. Describe what the word *equidistant* means in this situation.

# Perpendicular and Vertical

1. A **tangent** to a circle is a line that intersects the circle at exactly one point.

   The first diagram shows a circle with center $R$ and a line $l$ that is tangent to the circle at point $A$. In other words, $A$ is on both the circle and the line, and it is the only point on both.

   a. Explain how you can be sure $A$ is the point on line $l$ that is closest to point $R$. In other words, show that the segment from $R$ to $A$ is the shortest path from point $R$ to line $l$.

   b. Based on part a, what can you conclude about the relationship between $\overline{RA}$ and line $l$?

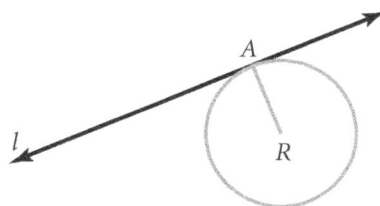

2. When two lines intersect, they form four angles. The angles formed by intersecting lines $l$ and $m$ in the diagram at the right are labeled 1, 2, 3, and 4.

   a. Suppose $\angle 1 = 20°$. Find the measures of the other three angles.

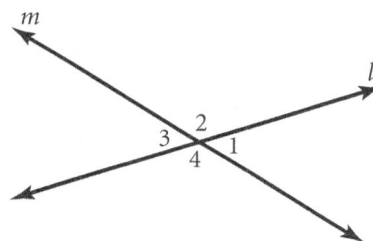

   Pairs of opposite angles, such as the pair $\angle 1$ and $\angle 3$ or the pair $\angle 2$ and $\angle 4$, are called **vertical angles.**

   b. Prove in general that vertical angles are equal. That is, show that $\angle 1 = \angle 3$ and $\angle 2 = \angle 4$. Don't use the value $20°$ for $\angle 1$, but you can use the reasoning by which you found the other angle measures in part a.

# On Patrol

You have learned that the distance from a point to a line is defined as the shortest possible distance from the point to the line. You have proved that this shortest path is perpendicular to the line. In this POW, the line is a highway and the point is a patrol station.

1. Two main highways intersect each other as they go through a certain county. The highway patrol wants to set up a station that will be the same distance from each highway. What are the possible choices for where to put the station? Assume the two highways are straight.

2. A neighboring county has three main highways. Assume all three highways are straight.

   a. Do all the highways have to intersect one another? What are the possibilities for how they might intersect? Draw some diagrams to illustrate the possibilities.

   b. Is it possible to place a patrol station the same distance from each of the three highways? How does the answer to this question depend on how the three highways intersect? Consider all possible arrangements of the highways and the possible locations of the station for each arrangement.

3. What about four highways? What about five? How can you generalize the result?

*continued*

1. *Problem Statement:* State the problem in mathematical language without reference to the context. That is, describe the problem in geometric terms without talking about patrol stations or highways.

2. *Process*

3. *Solution*

4. *Self-assessment*

Adapted from "Simple Math," a series of mathematics video-dramas produced by the Israeli Instructional Television Center, with academic adviser Nitsa Movshovitz-Hadar and producer Tamar Reiner.

# Proving with Distance—Part II

You showed in *Proving with Distance—Part I* that the points (5, 20), (9, 23), (12, 19), and (8, 16) are the vertices of a square.

The first diagram shows this square, as well as a circle that passes through, or intersects, the vertices of the square. We say that such a circle is **circumscribed** about the square or that the square is **inscribed** in the circle.

1. Find the center and radius of this circumscribed circle. Prove your answers are correct.

There is also a circle that passes through the midpoints of this square, as shown in the second diagram. The sides of the square are tangent to that circle. We say that such a circle is *inscribed* in the square or that the square is *circumscribed* about the circle.

2. Find the center and radius of this inscribed circle. Prove your answers are correct.

3. Is there a circumscribed circle for every quadrilateral? Justify your answer.

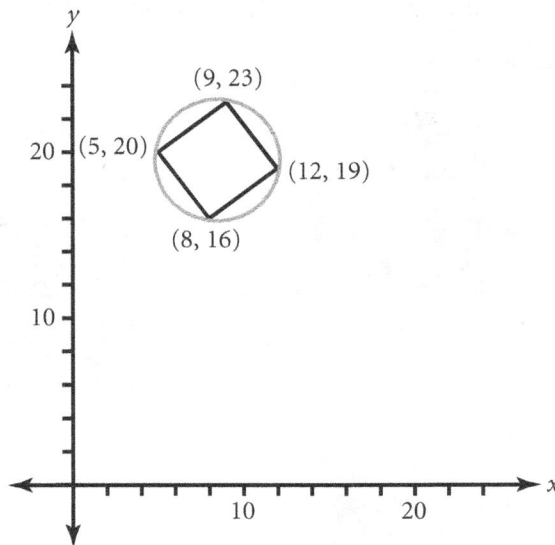

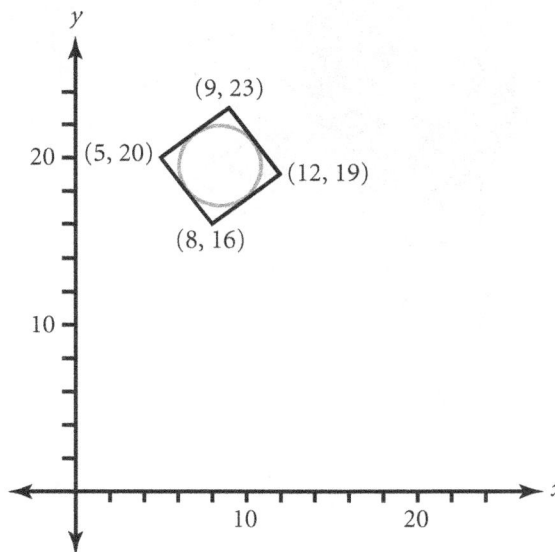

# All About Circles

You saw in *How Does Your Orchard Grow?* that Madie and Clyde need to understand how the circumference, area, and radius of a circle are related.

Your next task is to use polygons to help them understand the various relationships. What do circles have to do with polygons? You'll realize that the answer is as easy as pie—sort of.

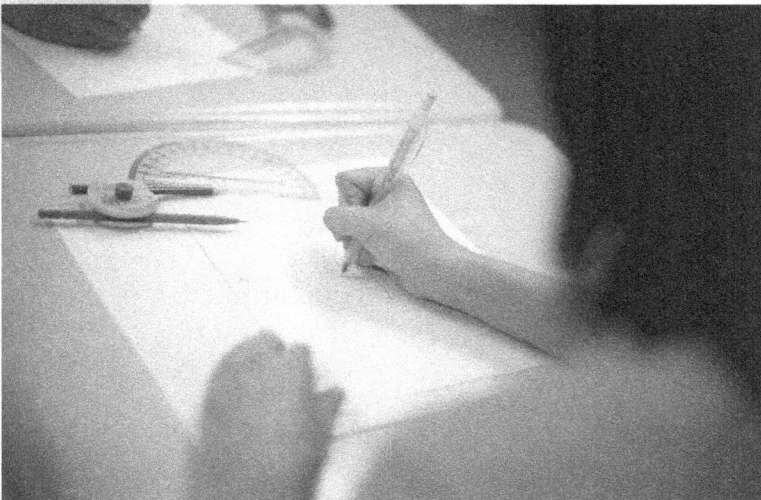

*A student uses a compass and straightedge to investigate the properties of a circle.*

# Squaring the Circle

Madie and Clyde need to know how the area and circumference of a circle are related to the radius. You will help them by completing a series of activities comparing a circle to various circumscribed regular polygons. In this first activity, you will compare the circle to the circumscribed square.

To draw such a figure, it's easiest to start by drawing a square on a sheet of grid paper. Then draw a circle inside the square so that it just touches the sides of the square. In other words, the circle is inscribed in the square.

These questions involve area, circumference, and perimeter. Find the requested numbers by making measurement estimates, rather than by using formulas.

1.  a. Estimate the area of the circle. That is, estimate the number of grid squares it contains.

    b. Estimate the area of the square.

    c. Find the ratio of your answers. That is, find the value of the fraction

$$\frac{\text{area of the circle}}{\text{area of the circumscribed square}}$$

*continued* ▶

2. a. Estimate the circumference of the circle. You may want to use string to estimate this length.

   b. Estimate the perimeter of the square.

   c. Find the ratio of your answers. That is, find the value of the fraction

   $$\frac{\text{circumference of the circle}}{\text{perimeter of the circumscribed square}}$$

3. Repeat Questions 1 and 2 for a larger circle and square. How do the ratios for the new circle-and-square pair compare to the ratios for the first pair?

4. Do you think the ratio you found in Question 1c will be the same for *any* circle? Write a careful explanation of your answer.

5. Do you think the ratio you found in Question 2c will be the same for *any* circle? Write a careful explanation of your answer.

# Using the Squared Circle

In *Squaring the Circle,* you compared circles to their circumscribed squares and found ratios involving area, circumference, and perimeter.

1. Give the estimates agreed on in class for each of these ratios.

    a. $\dfrac{\text{circumference of the circle}}{\text{perimeter of the circumscribed square}}$

    b. $\dfrac{\text{area of the circle}}{\text{area of the circumscribed square}}$

2. Consider the circle and circumscribed square shown here. The circle has a radius of 10 units.

    a. Find the length of a side of the square.

    b. Find the exact perimeter of the square.

    c. Combine your ratio from Question 1a with your answer to Question 2b to get an estimate of the circle's circumference.

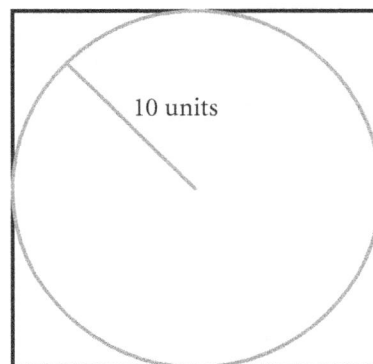

10 units

3. a. Examine the process you went through to answer Question 2c. Write a clear explanation of how to use the ratio from Question 1a to estimate the circumference of a circle from its radius.

    b. Use your explanation in Question 3a to write a formula, based on your ratio, that gives an estimate of a circle's circumference in terms of its radius.

4. a. Find the exact area of the square in Question 2.

    b. Use your answer from Question 4a and your ratio from Question 1b to estimate the circle's area.

5. a. Examine the process you went through to answer Question 4b. Write a clear explanation of how to use the ratio from Question 1b to estimate the area of a circle from its radius.

    b. Use your explanation in Question 5a to write a formula, based on your ratio, that gives an estimate of a circle's area in terms of its radius.

# Hexagoning the Circle

You found approximate formulas for the circumference and area of a circle of radius *r* by comparing the circle to its circumscribed square.

Now you will find formulas for the perimeter and area of the regular hexagon circumscribed about that circle. These values will provide fairly good estimates of the circumference and area of the circle.

1. Begin with a circle of radius 10. Find the perimeter and area of the regular hexagon circumscribed about this circle.

2. Generalize your work to a circle of radius *r*. Retrace your steps from Question 1 to create formulas for the perimeter and area of the circumscribed regular hexagon in terms of *r*.

# Octagoning the Circle

In *Hexagoning the Circle,* you found formulas for the perimeter and area of a regular hexagon circumscribed about a circle of radius $r$.

Your task now is to develop similar formulas for a circumscribed regular octagon, or eight-sided polygon.

1. Begin with a circle of radius 10. Find the perimeter and area of the regular octagon circumscribed about this circle.

2. Generalize your work to a circle of radius $r$. Retrace your steps from Question 1 to create formulas for the perimeter and area of the circumscribed regular octagon in terms of $r$.

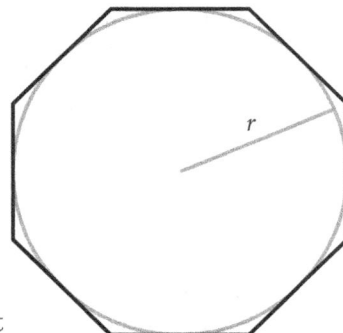

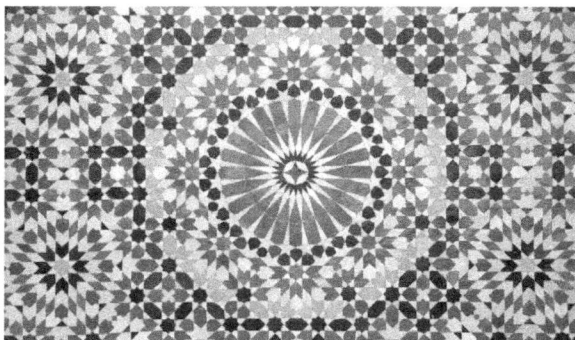

# Polygoning the Circle

The circumference $C$ and the area $A$ of a circle of radius $r$ can be found from formulas of the form

$$C = k_c r \quad \text{and} \quad A = k_a r^2$$

where $k_c$ and $k_a$ are specific numbers called **proportionality constants** or *constants of proportionality*.

Similar formulas exist for regular polygons circumscribed about a circle. You have already developed such formulas for a square, a regular hexagon, and a regular octagon. Comparing the circle to these circumscribed polygons can give estimates of the values of $k_c$ and $k_a$. The more sides a polygon has, the better the estimate.

Choose a number of sides (other than 4, 6, or 8), and develop formulas for the perimeter and area of the regular polygon with that many sides circumscribed about a circle of radius $r$.

# Another Kind of Bisector

In your work on the POW *Equally Wet,* you came across the idea of a perpendicular bisector. A *perpendicular bisector* is a line that splits a line segment into two equal parts. Line segments aren't the only geometric figures that have bisectors. An **angle bisector** is a ray that splits an angle into two equal parts. For example, in the diagram at the right, if ∠*RST* is 50° and ∠*RSU* and ∠*UST* each equal 25°, then ray $\overrightarrow{SU}$ is the bisector of ∠*RST*.

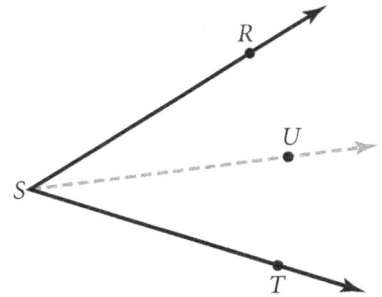

Suppose two lines *l* and *m* intersect at point *A*, as shown below. In this diagram, *B* and *D* are two points on *l*, and *C* and *E* are two points on *m*. The intersecting lines *l* and *m* form four angles at *A*: ∠*BAC*, ∠*CAD*, ∠*DAE*, and ∠*EAB*.

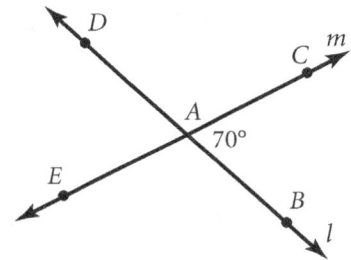

1. The diagram shows that ∠*BAC* is 70°. Find the measure of each of the other three angles.

2. In the third diagram, the dashed rays are the angle bisectors of the four angles. That is, $\overrightarrow{AP}$ bisects ∠*BAC*, $\overrightarrow{AQ}$ bisects ∠*CAD*, $\overrightarrow{AR}$ bisects ∠*DAE*, and $\overrightarrow{AS}$ bisects ∠*EAB*.

   a. Prove that the two angle bisectors $\overrightarrow{AP}$ and $\overrightarrow{AR}$ are part of the same line. That is, show that points *P*, *A*, and *R* are **collinear.** Use the fact that ∠*BAC* is 70° to show that ∠*PAR* is 180°.

   b. Prove that the angle bisectors $\overrightarrow{AQ}$ and $\overrightarrow{AS}$ are part of the same line.

   c. Prove that the line containing $\overrightarrow{AP}$ and $\overrightarrow{AR}$ is perpendicular to the line containing $\overrightarrow{AQ}$ and $\overrightarrow{AS}$.

3. Your work in Question 2 was based on the fact that ∠*BAC* is 70°. Now represent ∠*BAC* as *x* and prove Questions 2a, 2b, and 2c in general. You will have to express the other angles in terms of *x*.

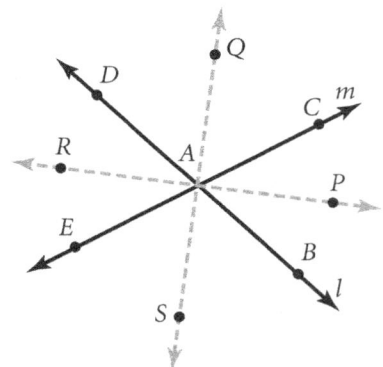

# Proving Triples

The set of numbers 3, 4, and 5 is called a **Pythagorean triple** because $3^2 + 4^2 = 5^2$. This numeric relationship means that a triangle with side lengths 3, 4, and 5 is a right triangle.

More generally, a Pythagorean triple is defined as any trio of positive whole numbers $a$, $b$, and $c$ that satisfy the equation $a^2 + b^2 = c^2$.

The numbers 4, 5, and 7 do not form a Pythagorean triple because $4^2 + 5^2$ does not equal $7^2$. This means that a triangle with sides of these lengths is not a right triangle.

1. Show that 5, 12, and 13 form a Pythagorean triple.

2. Suppose you multiply each member of a Pythagorean triple by the same positive number. For instance, suppose you multiply 3, 4, and 5 each by 6 to get 18, 24, and 30.

   a. Determine whether 18, 24, and 30 form a Pythagorean triple.

   b. Will the process of multiplying each member of a Pythagorean triple by the same number always produce another Pythagorean triple? Give at least two more examples that support your conclusion.

3. Prove the conclusion you reached in Question 2b. There are at least two possible approaches to this question. One involves similar triangles, and the other uses the distributive property.

# A Marching Strip

You might remember the economical king from *Eight Bags of Gold* and *Twelve Bags of Gold* in the Year 1 unit *The Pit and the Pendulum*. Well, he's back—with another weighty problem.

The king is planning a new rectangular courtyard in his palace. It will be laid out using square tiles. He has chosen some very pretty, but very inexpensive, tiles. However, his adviser has informed him that some of the tiles will have to be a more expensive type.

This is because visiting dignitaries will always walk along a certain diagonal, from one corner of the courtyard to the opposite corner, so the tiles along this diagonal will get lots of wear and tear. Every tile that contains a segment of the diagonal must thus be an extra-strength, more expensive tile. If a tile touches the diagonal only at a corner, it can be a regular tile.

The diagram shows what the situation would look like if the king made a 4-by-6 courtyard. In this case, 8 tiles (shaded darker than the rest) include some portion of the diagonal, so these would need to be the more expensive kind.

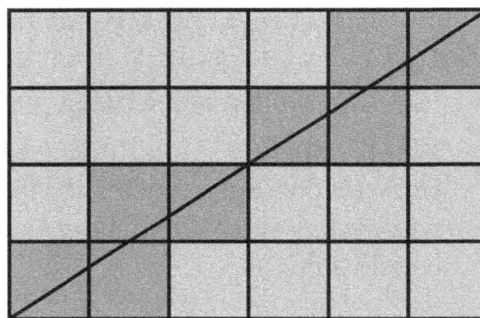

1. The king wants to know how many of these special tiles to order. At the moment, he is planning a 63-by-90 tile courtyard. That is, it will have 63 rows of tiles with 90 tiles in each row.

   If he stays with this plan, how many special tiles will he need?

2. The king keeps changing his mind about the dimensions of the courtyard. It would be especially helpful if you could find a general formula for him.

   Suppose the courtyard has $r$ rows with $c$ tiles in each row. How many special tiles will the king need? (The example illustrated in the diagram, which requires 8 special tiles, is the case $r = 4$ and $c = 6$.)

*continued*

○ *Write-up*

1. *Problem Statement*

2. *Process*

3. *Solution*

4. *Self-assessment*

Adapted with permission from *Mathematics Teacher,* by the National Council of
Teachers of Mathematics (May 1991).

# Orchard Growth Revisited

In *How Does Your Orchard Grow?* you were asked to describe a plan for solving the central unit problem based on certain information. That activity led to a search for formulas for the area and circumference of a circle in terms of its radius.

Now you have those formulas, and you can carry out the plan if you are given the necessary information about the trees. Here are two key facts about the trees in Madie and Clyde's orchard.

- The cross-sectional area of a tree trunk increases by 1.5 square inches per year.

- Right now, each tree trunk has a circumference of 2.5 inches.

Although you don't yet know the hideout tree radius, you can make progress toward solving the unit problem. Your task now is to answer this question.

*How long will it take for the trees to grow to a radius of 1 foot?*

# Cable Complications

While you've been busy learning about circles and coordinate formulas, Madie and Clyde have been dealing with another problem. There is an electrical cable that might get in the way of their planting.

While you solve this new problem, think about how it might be related to a key part of the unit problem—finding the hideout tree radius for an orchard of radius 50.

*Param Gii, Kimberly Lao, and Jessica Guzman work together to solve a problem involving an electrical cable.*

# Cable Ready

When Madie and Clyde bought their orchard, a straight electrical cable ran along the ground from the center of the orchard, at (0, 0) in their coordinate system, to the point (30, 20).

They wanted to start their planting while they waited for the electrical company to move the cable safely underground, but they had to be sure not to plant trees right on the cable.

1. a. Could they plant a complete mini-orchard of radius 1 at the center of their lot without planting right on the cable?

   b. Answer the same question for a mini-orchard of radius 2.

   c. What is the radius of the biggest complete mini-orchard Madie and Clyde could plant without planting on the cable? Assume the tree trunks are very thin.

2. Suppose Madie and Clyde plant that biggest possible mini-orchard from Question 1c. How big will the tree trunks have to become before one of them bumps into the cable?

With your group, prepare a poster that summarizes your work on Question 2 for presentation to the class.

# Going Around in Circles

1. At one time, Madie and Clyde had planned to plant a lawn instead of an orchard.

   Suppose the unit distance in their coordinate system is 10 feet. For instance, this would be the distance from $(0, 0)$ to $(1, 0)$.

   If a box of grass seed covers 300 square feet of ground, how many boxes would Madie and Clyde have needed for their entire lawn? Remember, the radius of their lot is 50 units, which you are assuming in this problem represents 500 feet.

2. A circular track has a diameter of 200 meters. How far is it around the track?

3. The distance around a circular pond is 100 feet. Will a 30-foot board be long enough to use as a bridge across the center of the pond?

4. A contractor has just finished installing some beautiful tiles on the floor of a circular room. Then he remembers that he is supposed to hang a banner in the room announcing the grand opening. The banner will reach across the room at its widest point, and the contractor needs to know this distance right away so he can order the banner. Unfortunately, he can't walk across the newly laid tile. He does know that he used 2830 square feet of tile for the floor.

   How long should the banner be?

# Daphne's Dance Floor

Daphne has a dream of building a huge dance floor in one of the meadows on her farm. Many families live in the area, but they have no place to go dancing. Daphne thinks that with a big-name band (like the Rocking Pebbles), she'll attract a full house.

She wants a circular dance floor, 100 feet across, which means a radius of 50 feet. She needs to order wood for the floor and for the railing to surround it, so she needs to find the area and circumference of a circle with a radius of 50 feet.

She doesn't know any formulas, but she has plenty of common sense. With her trusty compass, she makes a circle that has a radius of exactly 1 foot. Then she carefully wraps a string around the circle to measure the circumference.

1. What is the circumference of a circle with a radius of 1 foot?

Because the radius of her dance floor will be exactly 50 times the radius of the circle she drew, Daphne figures the circumference of the dance floor will be 50 times the circumference of her circle. She plans to use this value to calculate how much wood to order for the railing.

*continued* ▶

2. a. Multiply your answer to Question 1 by 50 to get Daphne's estimate of the circumference of the dance floor.

b. Use the formula for circumference to find the actual circumference of a circle with a radius of 50 feet.

c. Compare your answers to parts a and b. Explain whether Daphne's method works.

Next Daphne carefully cuts out some 1-inch squares and figures out how many of them it takes, including fractions, to fill in her small circle.

3. What is the area, in square inches, of a circle with a radius of 1 foot?

As her last step, Daphne multiplies this number by 50 and figures that this tells her how many square inches of wood flooring she will need to build her dance floor.

4. a. Multiply your answer to Question 3 by 50 to get Daphne's estimate.

b. Use the formula for area to get the actual area, in square inches, of a circle with a radius of 50 feet.

c. Explain where Daphne's plan went wrong, and describe what she should have done instead.

# Defining Circles

Earlier in this unit, you found that the equation $x^2 + y^2 = r^2$ describes the circle of radius $r$ with center at the origin $(0, 0)$. That is, points whose coordinates fit this equation are on the circle, and points that do not fit this equation are not on the circle.

Now you will generalize this formula to circles whose centers do not have to be at the origin.

1. Suppose the tree in the orchard at the point $(6, 2)$ is replaced by a sprinkler and the water reaches all points within 5 units of the sprinkler. Which trees will get wet?

2. Suppose a blade of grass is growing in the open space in the orchard at the point $(7.9, 6.1)$. Will this blade of grass get wet from the sprinkler in Question 1? How do you know?

3. Now suppose a blade of grass is growing in the open space in the orchard at the point $(x, y)$.

    a. How can you tell if this blade of grass will get wet from the sprinkler in Question 1?

    b. Write an equation for the points on the boundary of the region watered by the sprinkler.

4. Suppose the sprinkler is at $(a, b)$ and that the water from it reaches all points within $r$ units. Write an equation for the circle that forms the boundary of the region the sprinkler waters.

# The Standard Equation of a Circle

## The Equation of a Circle

According to the distance formula, the distance from $(x, y)$ to $(a, b)$ is given by the expression

$$\sqrt{(x - a)^2 + (y - b)^2}$$

This formula can be used to write the equation of a circle, because a circle is the set of points that are some fixed distance from a given point.

The equation $\sqrt{(x - a)^2 + (y - b)^2} = r$ says, in algebraic form, that the point $(x, y)$ is $r$ units from $(a, b)$. Thus the graph of this equation is the circle with center $(a, b)$ and radius $r$.

For convenience, we usually square both sides of this equation to avoid the square-root symbol. The standard form for the equation of this circle is

$$(x - a)^2 + (y - b)^2 = r^2$$

For example, the equation

$$(x - 8)^2 + (y + 5)^2 = 9$$

represents the circle with center $(8, -5)$ and radius 3, as shown here.

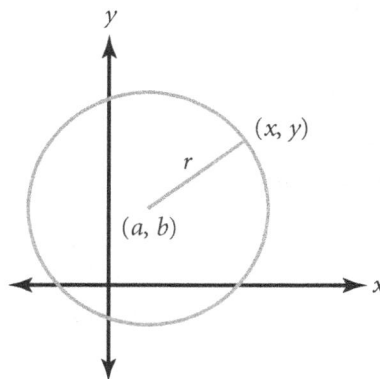

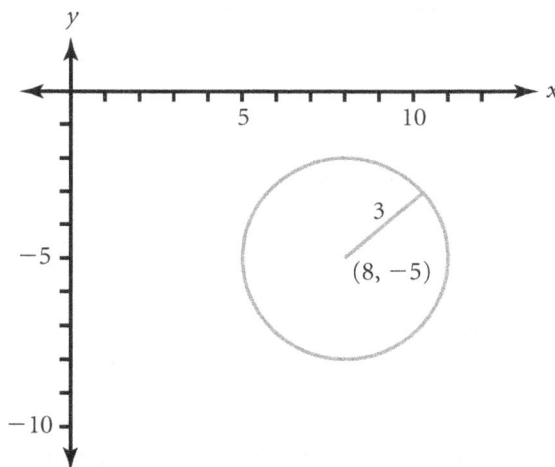

*continued* ▶

## Transforming the Equation

Multiplying out the terms of

$$(x - 8)^2 + (y + 5)^2 = 9$$

and simplifying the result gives the equivalent equation

$$x^2 - 16x + y^2 + 10y + 80 = 0$$

More generally, the equation of a circle is always equivalent to an equation of the form

$$x^2 + cx + y^2 + dy + e = 0$$

where the coefficients $c$, $d$, and $e$ depend on the center and radius of the circle.

## Reversing the Process

This process often can be done in reverse, going from an equation of the form

$$x^2 + cx + y^2 + dy + e = 0$$

to an equation with the standard form

$$(x - a)^2 + (y - b)^2 = r^2$$

Transforming the equation in this way, when it is possible, allows you to identify the center and radius of the graph simply by looking at the equation.

This process involves the technique called **completing the square.** For example, the expression $x^2 - 16x$ suggests the perfect square $(x - 8)^2$, because expanding $(x - 8)^2$ gives $x^2 - 16x + 64$.

# Completing the Square and Getting a Circle

Examine each equation. If possible, find an equivalent equation of the form

$$(x - a)^2 + (y - b)^2 = r^2$$

and identify the center and radius of the circle that the equation represents. If this is not possible, explain why not.

1. $x^2 - 8x + y^2 - 6y - 11 = 0$

2. $x^2 - 10x + y^2 + 12y + 28 = 0$

3. $x^2 + 3x + y^2 - 4y - 7 = 0$

4. $x^2 + 6x + y^2 + 2y + 13 = 0$

# Lines of Sight

In what direction should Madie and Clyde look to see out of their orchard for as long as possible? And how long will it take until their orchard becomes a hideout?

You're closing in on the solution to the unit problem (and the trees may be closing in on you!). You will need to combine many ideas from the unit to answer Madie and Clyde's questions.

*Alice Lenz uses models to help find the volume and lateral surface area of a cylinder.*

# The Other Gap

Using symmetry can make it much easier to locate the "last" line of sight from the center of the orchard, because some gaps between trees are like other gaps.

For instance, for an orchard of radius 3, all the gaps represented by the four shaded areas in the diagram are equivalent to the one in the first quadrant.

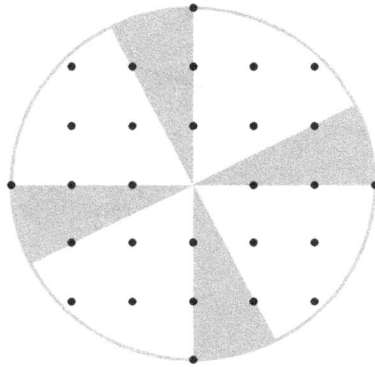

In fact, by symmetry, every gap is equivalent to one of the two gaps shown in the next diagram. The darkly shaded area represents the gap between the trees at points $A$ and $B$. The lightly shaded area represents the gap between the trees at points $B$ and $C$. Note that point $A$ is at $(1, 1)$, point $B$ is at $(2, 1)$, and point $C$ is at $(1, 0)$.

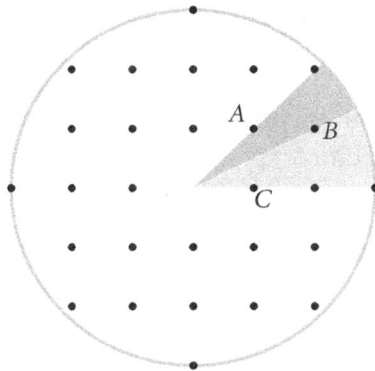

*continued*

Within each of these two gaps, there is one optimal line of sight—that is, a line of sight that stays unblocked as long as possible. Each is the optimal line in its gap because it goes through the midpoint of the segment connecting the two trees on either side of that gap. These optimal lines of sight are shown in the next diagram.

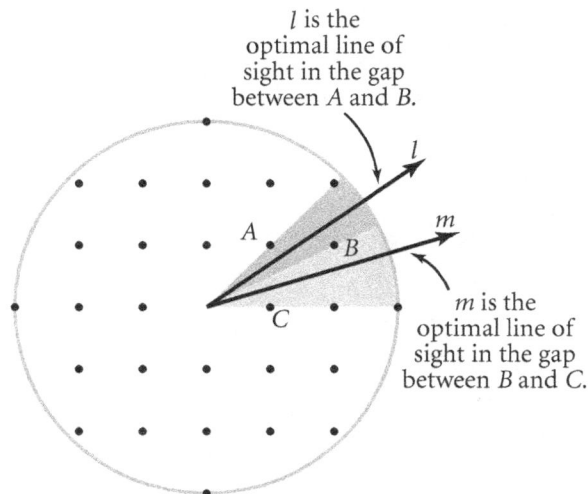

l is the optimal line of sight in the gap between A and B.

m is the optimal line of sight in the gap between B and C.

Note that in the context of this problem, lines of sight like l and m are really rays from the center of the orchard. For simplicity, we can think of them as lines.

Line l is in the gap between A and B. It goes through the midpoint of $\overline{AB}$, which is $(1\frac{1}{2}, 1)$. This line of sight follows the same path as the cable in the activity *Cable Ready*. You know from *Cable Ready* that this line of sight becomes blocked when the tree trunks reach a radius of about 0.28 unit.

The focus of this activity is on line *m*, which is the optimal line of sight in the gap between B and C. Line *m* goes through the midpoint of $\overline{BC}$, which is $(1\frac{1}{2}, \frac{1}{2})$. Your task is to answer this question.

*How big must the tree trunks become to block this line of sight?*

# Cylindrical Soda

A soft drink company sells its soda in cylindrical cans that are 12 centimeters tall and have a radius of 3 centimeters. A can of soda this size sells for 80¢.

1. Find the volume of this can (in cubic centimeters). Think about how the volume of a box is related to the area of its base, and apply similar reasoning to the **cylinder.**

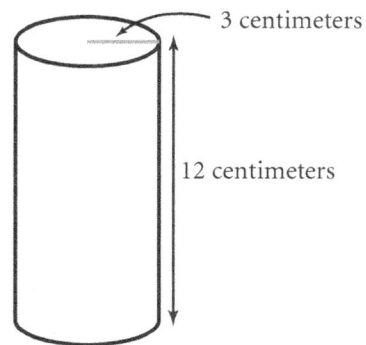

The company's Lateral Surface Department is responsible for purchasing metal for the curved surface of the cans. It does not buy the metal for the tops or bottoms of the cans.

3 centimeters

12 centimeters

2. How much metal should this department order per can (in square centimeters)? Assume there is no waste. Think about how the lateral surface area of a box is related to the perimeter of its base, and apply similar reasoning to the cylinder.

3. The company decides to make a new, taller container. This tall can will be twice the height of the standard-size can. The top and bottom will still be circles with radius 3 centimeters.

   a. If the company keeps the cost per ounce of soda the same, what should the selling price of the tall can be?

   b. How will the metal needed for the lateral surface area of the tall can compare to the metal needed for the standard can?

4. The company decides to make another new container. This can will be three times the width of the standard-size can. That is, the radius of the base will be 9 centimeters instead of 3 centimeters. The height of this wider can will be the standard 12 centimeters.

   a. If the company keeps the cost per ounce of soda the same, what should the selling price of the wider can be?

   b. How will the metal needed for the lateral surface area of the wider can compare to the metal needed for the standard can?

# Lines of Sight for Radius Six

The last diagram in the activity *The Other Gap* suggests that *m* is one of the last lines of sight to be blocked in an orchard of radius 3. The calculations in that activity confirm this suggestion. The relevant parts of that diagram are shown here.

Line of sight *m* goes through the gap between the tree at $B = (2, 1)$ and the tree at $C = (1, 0)$. It passes through the midpoint between these two trees, which is at $(1\frac{1}{2}, \frac{1}{2})$. By symmetry, there are other lines of sight that are just as good as *m*, but no line of sight is better than *m*.

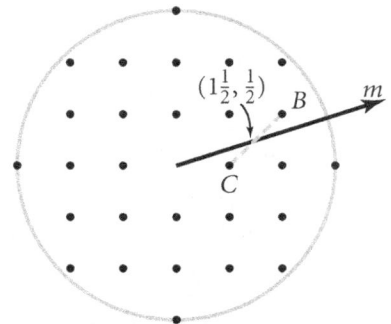

What are the last lines of sight for orchards of other sizes?

The next diagram shows the first quadrant of an orchard of radius 6. The center of the orchard is marked by a small dot, and shaded circles represent the trees.

Examine different lines of sight from the center of this orchard. Try to determine which of them will remain unblocked the longest.

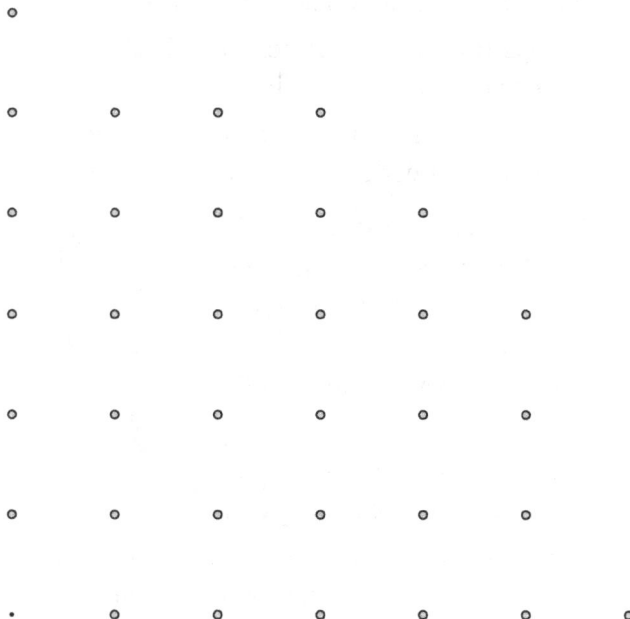

# Orchard Time for Radius Three

You know that for an orchard of radius 3, the line of sight shown in the diagram is one of the last to be blocked. The perpendicular distance from the tree at $(1, 0)$ to this line is $\frac{1}{\sqrt{10}}$ unit, or approximately 0.32 unit.

Recall these facts about Madie and Clyde's orchard from the activity *Orchard Growth Revisited*.

- The cross-sectional area of a tree trunk increases by 1.5 square inches per year.

- Right now, each tree trunk has a circumference of 2.5 inches.

Here is one additional piece of information about the orchard.

- The unit distance—for instance, the distance from $(0, 0)$ to $(1, 0)$—is 10 feet.

Your task is to answer this question.

*How long will it take for this line of sight to be blocked?*

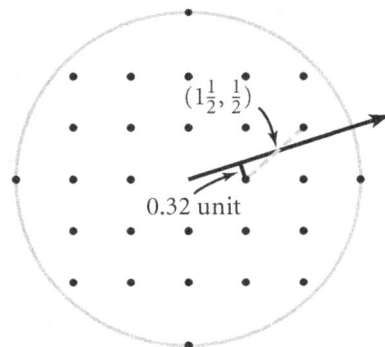

# Hiding in the Orchard

As you probably recall, Madie and Clyde planted their orchard on a circular lot of radius 50 units. They wonder how long it will take until they will no longer be able to see out of the orchard from the center.

You now have the information you need to solve their problem.

- The cross-sectional area of a tree trunk increases by 1.5 square inches per year.

- Right now, each tree trunk has a circumference of 2.5 inches.

- The unit distance—for instance, the distance from $(0, 0)$ to $(1, 0)$—is 10 feet.

- The last line of sight is the line that goes from the origin through the point $(25, \frac{1}{2})$.

Your task is to put all of this information together to answer this question.

*How soon after they planted the orchard will the center of the lot become a true orchard hideout?*

# Big Earth, Little Earth

Suppose you were able to wrap a very long piece of string tightly around the equator of the earth. And suppose that at the same time, your friend were to wrap a piece of string tightly around the equator of a classroom globe. Of course, your friend's string would be much shorter.

For simplicity, assume the earth is a perfect sphere with a radius of 4000 miles (this isn't far off) and that the globe is a perfect sphere with a radius of 6 inches. Remember that 1 mile is equal to 5280 feet.

1. Find the length of each piece of string.

Now suppose you each replace your string with a piece 1 foot longer. Of course, your strings will no longer fit tightly around their equators. But if the earth's radius got a little bit bigger, your string would be tight again. And if the globe's radius grew a little bit, your friend's string would be tight again.

2. By how much would each radius have to increase? Which radius would have to grow more—the earth's or the globe's—for the string to fit tightly? Justify your answer.

# Beginning Portfolios

## A Coordinate Summary

Many of the ideas in this unit involve the coordinate system. For example, you developed a formula for finding the distance between two points in the coordinate plane in terms of their coordinates.

Write a summary of the main ideas about coordinates that you learned or used in this unit. Include important formulas and equations and their explanations as well as examples of how coordinates played a role in solving the unit problem. Also indicate which activities played an important role in your understanding of these ideas.

## Me, Myself, and Pi

Maybe you knew something about the number $\pi$ before starting this unit. Whether you did or not, your work in this unit with circles and circumscribed polygons will have given you insight into this important number.

Write an essay on what you thought about $\pi$ before this unit and what you think now. Include any misconceptions you had that you have now corrected. In your essay, indicate which activities played an important role in your understanding of $\pi$.

# Orchard Hideout Portfolio

You will now put together your portfolio for *Orchard Hideout*. This process has three steps.

- Write a cover letter that summarizes the unit.
- Choose papers to include from your work in the unit.
- Discuss your personal growth during the unit, especially with regard to the relationship between geometry and algebra.

## Cover Letter

Look back over *Orchard Hideout* and describe the central problem of the unit and the key mathematical ideas. Your description should give an overview of how the key ideas were developed and how they were used to solve the central problem.

In compiling your portfolio, you will select some activities you think were important in developing the unit's key ideas. Your cover letter should include an explanation of why you selected each item.

## Selecting Papers

Your portfolio for *Orchard Hideout* should contain these items.

- *Hiding in the Orchard*
- A Problem of the Week

  Select any one of the three POWs you completed in this unit: *Equally Wet, On Patrol,* or *A Marching Strip.*

- *Beginning Portfolios*

  Include the write-up you did for this activity as well as the activities you discussed in it.

*Arta Gharib Parsa selects work for his portfolio.*

*continued*

- Other key activities

  Identify two concepts you think were important in this unit (other than those discussed in *Beginning Portfolios*). For each concept, choose one or two activities that improved your understanding, and explain how the activities helped.

## Personal Growth

Your cover letter for *Orchard Hideout* describes how the mathematical ideas developed in the unit. In addition, write about your own personal development during this unit. You may want to address this question.

*How did the unit improve your understanding of the relationship between algebra and geometry?*

Include any thoughts about your experiences that you wish to share with a reader of your portfolio.

# SUPPLEMENTAL ACTIVITIES

The supplemental activities in *Orchard Hideout* expand the geometry themes from the unit. Here are some examples.

- *The Perpendicular Bisector Converse* and *Why Do They Always Meet?* follow up on ideas from the POW *Equally Wet*.

- *Inscribed Angles* and *Angles In and Out* are among several activities involving angles and angle measurement.

- *Right in the Center* and *Hypotenuse Median* are related activities involving coordinates.

# Right and Isosceles

Right triangles are special, and so are isosceles triangles. Some triangles have the special property of fitting into both categories.

Triangle $ABC$ has a right angle at $C$, and $\overline{AC}$ and $\overline{BC}$ have the same length.

1. What is the size, or measure, of $\angle A$? Justify your answer.

2. Find the exact values of $\sin A$, $\cos A$, and $\tan A$, based on geometric principles. In other words, do not use a calculator to get your answers.

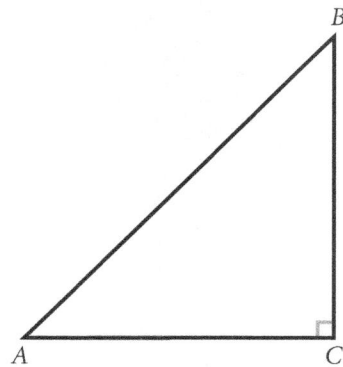

# The Perpendicular Bisector Converse

In the POW *Equally Wet,* you investigated where a sprinkler could be placed to water two flowers equally. In mathematical terms, this means finding the set of points in a plane that are equidistant from two given points. You saw that this set forms the perpendicular bisector of the segment connecting the two points.

One part of this result involved proving this "if-then" statement, which you did in the activity *A Perpendicularity Proof.*

*If D is on the perpendicular bisector of the segment connecting points A and B, then the distances AD and BD are equal.*

In this statement, the hypothesis is "*D* is on the perpendicular bisector of the segment connecting points *A* and *B.*" The conclusion is "the distances *AD* and *BD* are equal."

When the hypothesis and conclusion of an "if-then" statement are interchanged, the result is called the **converse** of the original statement. For some true "if-then" statements, the converse is true. For others, the converse is false.

*continued* ◗

## Your Task

Your task is to prove that the converse of the statement you proved earlier is also true. That is, prove this statement.

*If the distances AD and BD are equal, then D is on the perpendicular bisector of the segment connecting points A and B.*

Start with a diagram like the one here, and assume the hypothesis is true. That is, assume $AD = BD$.

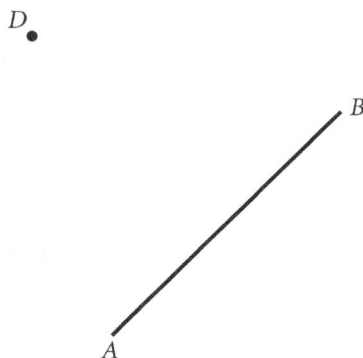

Then draw a perpendicular segment from $D$ to $\overline{AB}$, as in the next diagram.

At this point you know that $\overline{DC}$ is perpendicular to $\overline{AB}$, but you don't know whether it bisects $\overline{AB}$. You need to show that $C$ is the midpoint of $\overline{AB}$. In other words, show that $AC = BC$. The key is finding a way to use the hypothesis that $AD$ and $BD$ are equal.

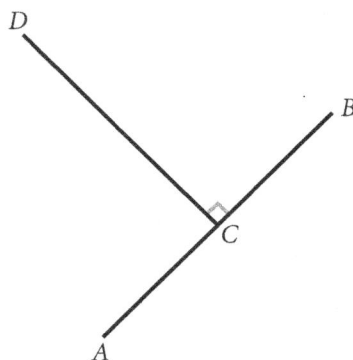

# Counting Trees

If Madie and Clyde plant their orchard on a lot of radius 1, it will include only the four trees on the axes. All four of those trees lie on the boundary of the orchard.

If they expand to a lot of radius 2, their orchard will include 12 trees. Again, there will be exactly four trees on the boundary.

1. Sketch an orchard of radius 3. Find out how many trees it has on its boundary.

2. Sketch an orchard of radius 5. Find out how many trees it has on its boundary.

3. Is it possible for an orchard to have on its boundary a tree whose coordinates are equal? Assume the radius of the lot is a positive integer.

4. Use your examples and your answer to Question 3 to develop a general statement about the number of trees on the boundary of an orchard. You may find your work on the activity *Other Trees* useful.

# Perpendicular Bisectors by Algebra

The perpendicular bisector of a line segment is the set of points that are equidistant from the endpoints of the segment. The coordinate system provides a way to think about perpendicular bisectors algebraically.

1. Start with the two points (4, 3) and (8, 5) and a general point $(x, y)$.

    a. Write an expression for the distance from $(x, y)$ to (4, 3).

    b. Write an expression for the distance from $(x, y)$ to (8, 5).

    c. Form an equation by setting the two expressions equal.

Your equation from Question 1c says, in effect, "$(x, y)$ is equidistant from (4, 3) and (8, 5)." This means it should be the equation of the perpendicular bisector of the segment connecting those two points.

2. Simplify your equation as much as possible. You should be able to simplify it so that it becomes a linear equation.

3. a. Find the coordinates of the midpoint of the segment connecting (4, 3) and (8, 5).

    b. Verify that the coordinates you found fit the equation from Question 2.

    c. Explain why the coordinates of the midpoint *should* fit the equation.

4. Plot the points (4, 3) and (8, 5) and draw the segment connecting them. Then graph your equation from Question 2. Does the result fit your expectations? Explain.

5. Suppose $(a, b)$ and $(c, d)$ are any two points. Use the steps in Questions 1 and 2 to write the equation of the set of points equidistant from these two points. That is, find a linear equation for the perpendicular bisector of the segment connecting $(a, b)$ and $(c, d)$.

# Midpoint Proof

In the activity *A Snack in the Middle,* you were asked to find the midpoint of the segment connecting (24, 6) and (30, 14). The answer is (27, 10). Part of your task was to prove that (27, 10) is equidistant from the two endpoints of the segment.

Unfortunately, this doesn't prove that (27, 10) is the midpoint, because *any* point on the perpendicular bisector of the segment will be equidistant from the two endpoints.

1. Find at least two other points that are equidistant from (24, 6) and (30, 14).

2. Explain how you know that (27, 10) is actually on the line segment connecting (24, 6) and (30, 14). To do this, you will need to find a linear equation whose graph includes those two points, and show that (27, 10) also fits the equation.

3. How would you prove in general that the midpoint formula actually gives the midpoint?

# Why Do They Always Meet?

Start with any triangle. Draw the perpendicular bisector of one of the triangle's sides. Then draw the perpendicular bisector of a different side.

Because these two perpendicular bisectors aren't parallel, they must meet at some point.

1. Prove that the third perpendicular bisector must go through this same point. In other words, show that all three perpendicular bisectors meet in a single point. Use the principle that the perpendicular bisector of a line segment consists of all points equidistant from the endpoints of the segment.

2. Explain how your result from Question 1 shows that every triangle has a circumscribed circle.

3. Explain how this idea applies to the three-flower case in the POW *Equally Wet*.

4. If you didn't answer the sprinkler problem for four flowers or more in your POW, work on it some more, using the ideas from Questions 1 to 3.

# Inscribed Angles

In examining the three-flower case in the POW *Equally Wet,* you may have realized that any triangle can be inscribed in a circle. You can find the center of this circle by looking for the place where the perpendicular bisectors of the three sides of the triangle meet. That point is equidistant from all three vertices. (The supplemental activity *Why Do They Always Meet?* asks you to prove that the three perpendicular bisectors all meet in one point.)

Now you will look at the special case in which the center of the circle is actually on one of the sides of the triangle.

In this diagram, $\triangle ABC$ is inscribed in a circle whose center, point $D$, is on side $\overline{AB}$. Because $\overline{AB}$ is a diameter of the circle, we say that $\triangle ACB$ is *inscribed in a semicircle.* The diagram also shows the radius from $D$ to $C$ and uses the letters $u$, $v$, $w$, $x$, $y$, and $z$ to represent the measures of the indicated angles. For example, $u$ is the measure of $\angle ADC$.

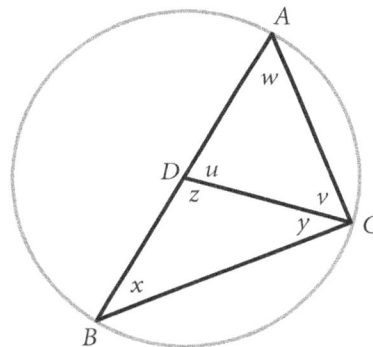

Your task is to prove this statement.

> *An angle inscribed in a semicircle must be a right angle.*

You can use these steps to construct a proof.

1. Explain why $x = y$ and $v = w$.

2. Explain why $u + z = 180°$.

3. Write equations showing the sums of the angles of various triangles.

4. Combine your findings from Questions 1 to 3 to prove that $v + y = 90°$. In other words, prove that $\angle ACB$ is a right angle.

# More Inscribed Angles

The supplemental activity *Inscribed Angles* asked you to prove that an angle inscribed in a semicircle must be a right angle. This is actually a special case of a more general principle. To state this general principle, we need to introduce some terminology.

If an angle *ABC* is situated with points *A*, *B*, and *C* on a circle, as shown at the right, we say the angle is *inscribed in the circle*.

The angle formed by the radii from the center *D* to points *A* and *C*, which is ∠*ADC*, is called the **central angle** that corresponds to inscribed ∠*ABC*.

Here is the general principle.

> *An angle inscribed in a circle is half the size of the corresponding central angle.*

Before proving the general principle, you need to consider first the case in which $\overline{AB}$ is actually a diameter of the circle, as shown in the second diagram. (This diagram is part of the diagram from the activity *Inscribed Angles*.)

1. Prove in this case that ∠*ABC* is half the size of ∠*ADC*. (Look at the relevant parts of Questions 1 to 3 of *Inscribed Angles*.)

2. Apply the special case from Question 1 to prove the general case. You will need to show that ∠*ABE* in the third diagram is half the size of ∠*ADE* and that ∠*CBE* is half the size of ∠*CDE*.

3. Explain why the result from *Inscribed Angles* is a special case of this principle.

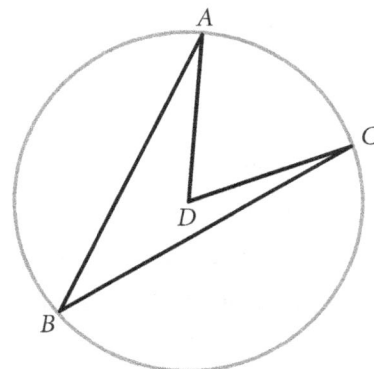

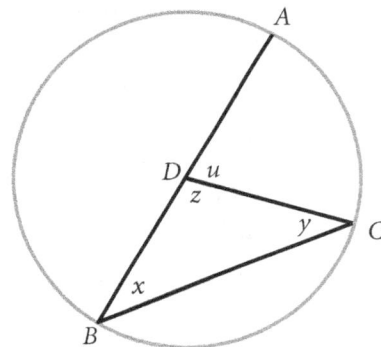

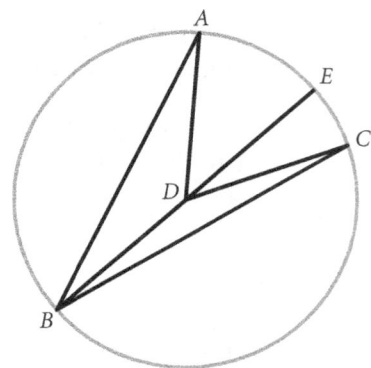

# Angles In and Out

If you worked on either of the supplemental activities *Inscribed Angles* or *More Inscribed Angles*, you saw a diagram like this one.

In the course of your work, you may have realized that $u = x + y$. In fact, this is a special case of a more general principle that has nothing to do with circles or inscribed angles.

Consider the second diagram, in which $\triangle PQR$ is an arbitrary triangle and $S$ is a point on the extension of $\overline{PQ}$. The letters $a$, $b$, $c$, and $d$ represent the measures of the indicated angles.

You may recall that an angle such as $\angle SQR$ is called an **exterior angle** for $\triangle PQR$. In addition, $\angle PQR$ is called the **adjacent interior angle** for this exterior angle, and $\angle QPR$ and $\angle QRP$ are called the **nonadjacent interior angles.**

Your task in this activity is to prove this statement.

*In any triangle, each exterior angle is equal to the sum of the two nonadjacent interior angles.*

In other words, prove that $d = b + c$.

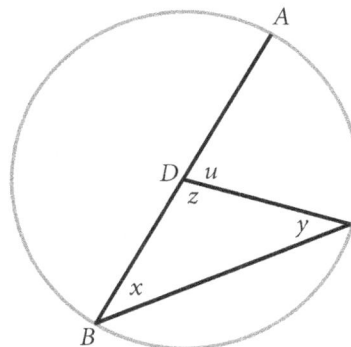

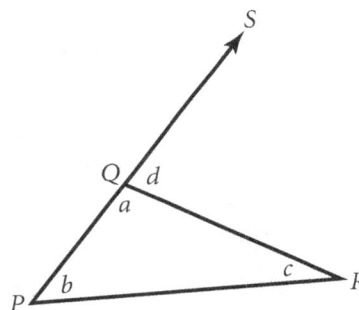

# Midpoint Quadrilaterals

In Question 2 of the activity *Proving with Distance—Part I,* you picked four points in the coordinate plane and connected them to form a quadrilateral.

You then found the midpoints of the sides of that quadrilateral and connected them to form a new quadrilateral, which we will call the *midpoint quadrilateral.*

Your final diagram probably looked something like the pair of quadrilaterals shown here.

It turns out, as you may have conjectured, that the midpoint quadrilateral is always a parallelogram, no matter what quadrilateral you start with. Your task now is to prove this. (In *Proving with Distance—Part I,* you were asked to prove that the opposite sides of that specific quadrilateral had equal lengths.)

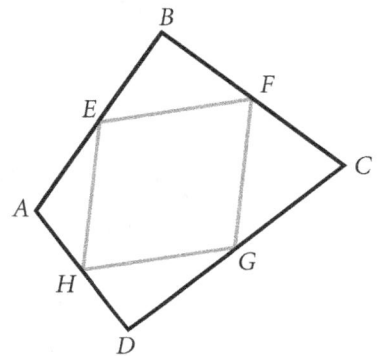

Before beginning, recall that a *parallelogram* is defined as a quadrilateral with opposite sides that are parallel.

## An Outline Using Triangles

This activity provides a three-part outline to help you construct a proof. (Parts I and II involve only triangles.)

The diagram at the right is formed from part of the original diagram and shows the diagonal $\overline{AC}$ of the original quadrilateral. So $\triangle ABC$ could be any triangle, and $E$ and $F$ are the midpoints of $\overline{AB}$ and $\overline{BC}$.

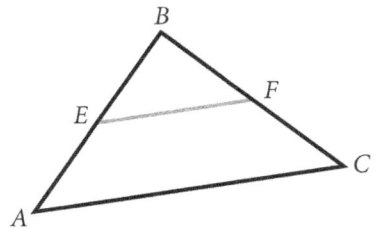

In $\triangle ABC$, we will refer to $\overline{AC}$ as the *base* and to $\overline{EF}$ as the **midline** (because it connects two midpoints).

*continued* ▶

In Parts I and II, you will prove two statements about this diagram.

- *EF* is half *AC*. (In other words, the length of $\overline{EF}$ is half the length of $\overline{AC}$.)
- $\overline{EF}$ is parallel to $\overline{AC}$.

You may want to think ahead about how you would use the second statement to prove *EFGH* is a parallelogram. (That's Part III.) Keep in mind that these statements apply to any triangle.

## Part I: The Length of the Midline Is Half the Length of the Base

The proof of this statement suggested in Questions 1 and 2 uses coordinates. (The study of geometry using coordinates is called **analytic geometry.**)

To make things completely general, suppose that point *A* is $(r, s)$, *B* is $(t, u)$, and *C* is $(v, w)$.

1. Find expressions for *EF* and *AC* in terms of these coordinates.

2. Use your expressions from Question 1 to prove that *EF* is half *AC*.

## Part II: The Midline Is Parallel to the Base

The proof of this statement suggested in Questions 3 and 4 uses similarity but does not involve coordinates. (The more general study of geometry, without coordinates, is sometimes called **synthetic geometry.**)

3. Prove that $\triangle EBF$ is similar to $\triangle ABC$. Use the result from Question 2 to show that corresponding sides of these two triangles are proportional.

4. Prove that $\overline{EF}$ is parallel to $\overline{AC}$. Use the result from Question 3 to show that $\angle BEF = \angle BAC$. Then use ideas about transversals and their corresponding angles.

## Part III: EFGH Is a Parallelogram

5. Prove that quadrilateral *EFGH* is a parallelogram. Use the fact that for any triangle, the midline is parallel to the base. Also use the definition of a parallelogram.

# Equidistant Lines

As part of your discussion of the activity *Down the Garden Path,* you proved this statement.

*If a line goes through the midpoint of $\overline{AB}$, then it is equidistant from points A and B.*

Your task now is to prove the converse of this statement, which is the conjecture the class came up with by drawing various lines in *Down the Garden Path.* In other words, prove this statement.

*If a line is equidistant from points A and B, then it goes through the midpoint of $\overline{AB}$.*

You can use the diagram here to set up the problem. The diagram shows two points *A* and *B* and a line *l*. Point *C* is where *l* meets $\overline{AB}$.

The hypothesis says to assume that *l* is equidistant from *A* and *B*. This means the perpendicular distances *AD* and *BE* are equal.

You need to prove that *C* is the midpoint of $\overline{AB}$.

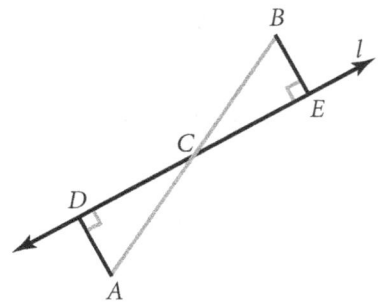

# Right in the Center

1. Consider the triangle whose vertices are (4, 1), (12, 7), and (24, −9).

   a. Prove this is a right triangle.

   b. Find the midpoint of the hypotenuse of this triangle.

   c. Show that the midpoint of the hypotenuse is equidistant from all three vertices.

2. Use the results from Question 1 to find the center and radius of a circle that goes through all three vertices of the triangle.

3. Find the equation for the circle in Question 2. Verify that the coordinates of the points (4, 1), (12, 7), and (24, −9) all fit the equation. (This requires ideas from the activity *Defining Circles*.)

# Thirty-Sixty-Ninety

Triangle *ABC* in this diagram is equilateral, and $\overline{AD}$ is the altitude from *A* to $\overline{BC}$.

The supplemental activity *Right and Isosceles* focused on one special type of triangle. The focus now is on the equilateral triangle.

1. Find the angle measures in right triangle *ABD*.

2. Use the diagram to find the exact values of sin 30°, cos 30°, tan 30°, sin 60°, cos 60°, and tan 60°. Do not rely on a calculator for your answers.

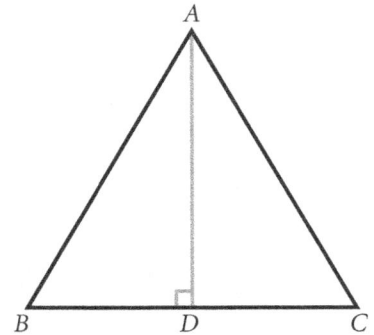

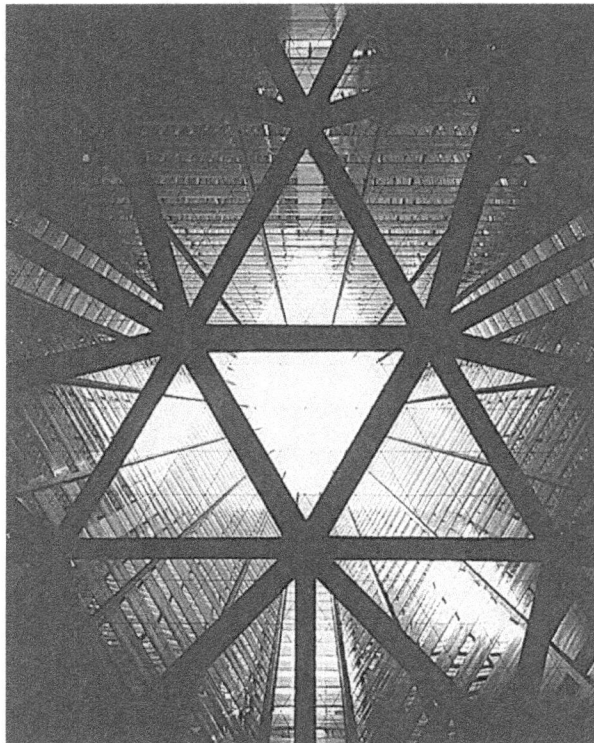

# Darts

To pass the time while their trees are growing, Madie and Clyde hang a dartboard on one of the trees. The dartboard looks like the one shown here, in which the radii of the circles are in the ratio 1:2:3.

A dart that lands in the white ring around the outside earns 10 points. A dart in the middle shaded ring earns 30 points. A dart that hits the innermost circle earns 50 points.

Assume that Madie and Clyde always hit the board and that darts have an equal chance of landing at any location on the board. What is the expected value for their score per dart? (*Reminder:* You can find the expected value by taking an average over a large number of trials.)

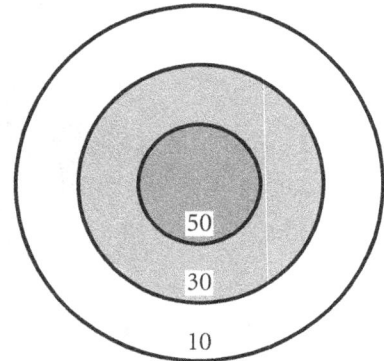

Adapted with permission from the *Mathematics Teacher,* by the National Council of Teachers of Mathematics (December 1990).

# More About Triples

As you have seen, a Pythagorean triple is any trio of positive whole numbers $a$, $b$, and $c$ that satisfy the equation $a^2 + b^2 = c^2$. The triple 3, 4, and 5 and the triple 5, 12, and 13 are two of the most familiar examples.

Your task in this activity is to learn more about Pythagorean triples. Here are two questions to get you started, but don't be limited by them. In fact, the study of problems like these is part of a branch of mathematics called *number theory*. You may want to consult a book on number theory for ideas.

1. Is it possible to have a Pythagorean triple $a$, $b$, and $c$ in which $a$ and $b$ are equal? Justify your answer.

A *primitive* Pythagorean triple is one in which $a$, $b$, and $c$ have no common whole-number factor (other than 1).

2. Assume $a$, $b$, and $c$ form a primitive Pythagorean triple, with $a^2 + b^2 = c^2$.

   a. Prove that $a$ and $b$ cannot both be even numbers.

   b. Prove that $a$ and $b$ cannot both be odd numbers. (This is harder than part a.)

   c. Prove that $c$ must be an odd number.

# The Inscribed Circle

You saw in the POW *On Patrol* that a point on an angle bisector is equidistant from the two lines that form that angle.

This fact and its converse can be used to prove that every triangle has an inscribed circle. (This is similar to the result from Question 2 of the supplemental activity *Why Do They Always Meet?*) An *inscribed circle* is a circle to which the sides of the triangle are all tangent.

Consider $\triangle ABC$, shown here. Suppose rays $l$ and $m$ are the bisectors of $\angle BAC$ and $\angle BCA$. Point $D$ is where these two rays meet.

1. Prove that the ray from $B$ through $D$ bisects $\angle ABC$. Use the general principle about angle bisectors from *On Patrol*.

2. Explain how your result proves that every triangle has an inscribed circle.

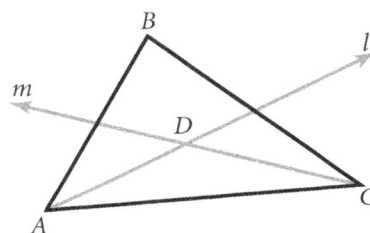

# Medians and Altitudes

If you worked on the supplemental activity *Why Do They Always Meet?*, you saw why the perpendicular bisectors of the sides of a triangle all meet in a single point. That point is called the *circumcenter* of the triangle, because it is the center of the circumscribed circle. The diagram at the right shows a triangle, its perpendicular bisectors, and the circumscribed circle.

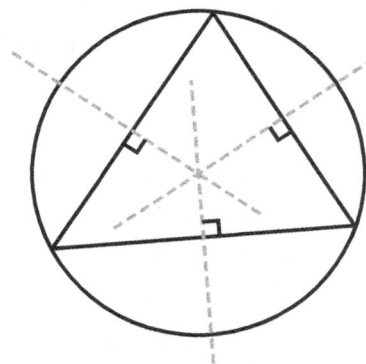

A similar principle holds for angle bisectors. That is, the three angle bisectors of a triangle all meet in a single point. That point is called the *incenter* of the triangle, because it is the center of the inscribed circle. (This principle was established in the supplemental activity *The Inscribed Circle*.)

This activity involves two other, similar principles. The proofs of these two principles are much more difficult, so this is a very challenging activity.

## Part I: Medians

In any triangle, the line segment connecting a vertex to the midpoint of the opposite side is called a **median.** For example, in $\triangle RST$, point $W$ is the midpoint of $\overline{RS}$, so $\overline{TW}$ is a median.

Prove that in any triangle, the three medians meet in a single point.

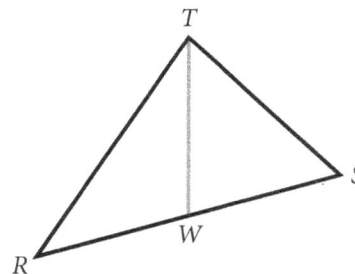

*continued* ▶

## Part II: Altitudes

In any triangle, the line segment from a vertex perpendicular to the line containing the opposite side is called an *altitude*. For example, in $\triangle FGH$, $\overline{GK}$ is perpendicular to $\overline{FH}$, so $\overline{GK}$ is an altitude.

Prove that in any triangle, the three altitudes meet in a single point.

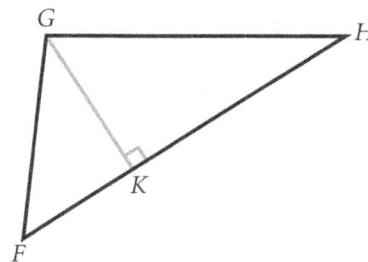

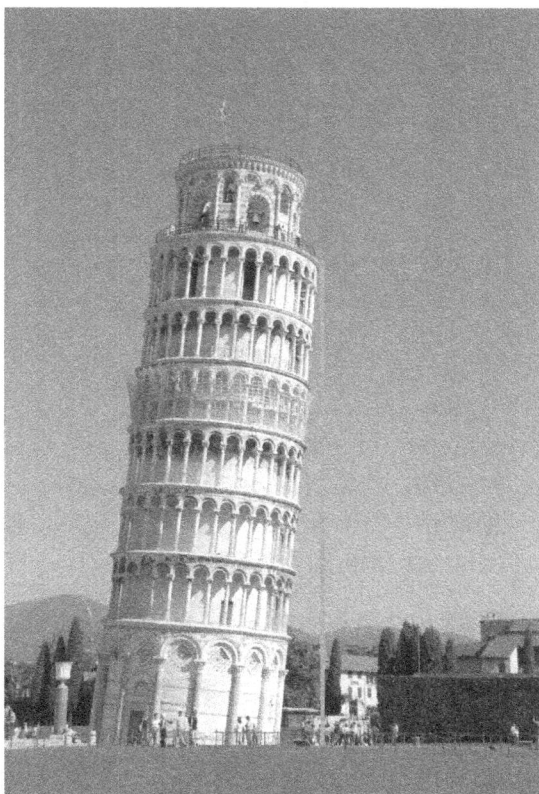

# Hypotenuse Median

In the supplemental activity *Right in the Center,* you were asked to show that the midpoint of the hypotenuse of a certain right triangle is equidistant from all three vertices. As you may have guessed, this is not a coincidence. Your task now is to prove the general principle.

In any triangle, the line segment connecting a vertex to the midpoint of the opposite side is called a *median.* So, $\overline{CD}$ in the diagram is the median to the hypotenuse. In this diagram, $\triangle ABC$ is a right triangle and $D$ is the midpoint of $\overline{AB}$.

The goal of this activity is to prove this principle.

> *The length of the median to the hypotenuse is half the length of the hypotenuse.*

We will call this principle the *median-to-the-hypotenuse property.*

Here are outlines for two ways to prove this principle. Try to prove it both ways.

1. First prove that the two diagonals of any rectangle are the same length and bisect each other. Draw $\triangle ABC$, draw in the full rectangle around $\triangle ABC$, and use this fact about the diagonals of a rectangle to prove the median-to-the-hypotenuse property.

2. Place the triangle in the coordinate system with point $C$ at the origin and points $A$ and $B$ on the axes. Assign coordinates to $A$ and $B$ in a way that makes $\triangle ABC$ a general right triangle. Then find the coordinates of $D$ and use the result to prove the median-to-the-hypotenuse property.

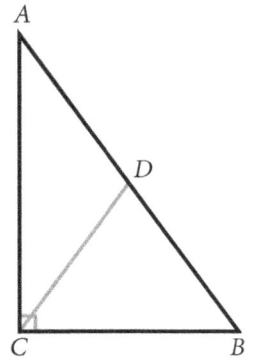

# Not Quite a Circle

You've seen that $x^2 + y^2 = r^2$ is the equation of a circle with center $(0, 0)$ and radius $r$. A point with coordinates $(x, y)$ is on this circle if and only if the coordinates fit the equation.

Your task in this activity is to experiment with variations of this equation to determine what shapes they define.

Specifically, think of the equation in the form $1 \cdot x^2 + 1 \cdot y^2 = r^2$, and examine what happens to the graph of the equation if numbers other than 1 are used as coefficients for $x^2$ and $y^2$. That is, consider equations such as these.

$$9x^2 + 16y^2 = 144$$
$$25x^2 + 16y^2 = 400$$
$$4x^2 + 4y^2 = 25$$

Restrict yourself to coefficients and **constant terms** that are perfect squares, as in these examples. This will allow you to write these numbers as squares of other numbers. For example, the equation $9x^2 + 16y^2 = 144$ can be written $3^2 \cdot x^2 + 4^2 \cdot y^2 = 12^2$.

Investigate specific examples and report on what you learn about the graphs of equations of the form $a^2x^2 + b^2y^2 = c^2$.

# Knitting

As the years go by, Madie and Clyde decide to start a family. They spend their afternoons sitting out among the trees, knitting booties and doing geometry problems.

They figure that a ball of yarn with about a 3-inch diameter will be needed for each bootie.

1. Madie has a ball of yarn with a 1-foot diameter. How many booties will it yield?

2. How many booties will they get from a ball of yarn with a 2-foot diameter? How does this compare with your answer to Question 1?

Adapted with permission from *Mathematics Teacher,* by the National Council of Teachers of Mathematics (December 1990).

# What's a Parabola?

In the Year 2 unit *Fireworks*, you described graphs for $y = x^2$ and similar quadratic functions as parabolas. But what exactly is a parabola? Is it any curve that goes down and then up (or up and then down)?

In fact, this term has a precise geometric meaning that can be described in two distinct ways.

## The Parabola as a Conic Section

The **parabola** is one of several shapes, called *conic sections,* that you can get by taking the intersection of a plane with a cone. In mathematics, a cone consists of two parts, sort of like one ice cream cone inverted on top of another and extending infinitely in both directions.

By changing the angle of the plane, you get different conic sections. These include the familiar shape known as an **ellipse** (which includes the circle as a special case) and another shape called a **hyperbola.** When the plane is parallel to the edge of the cone, the result is a parabola.

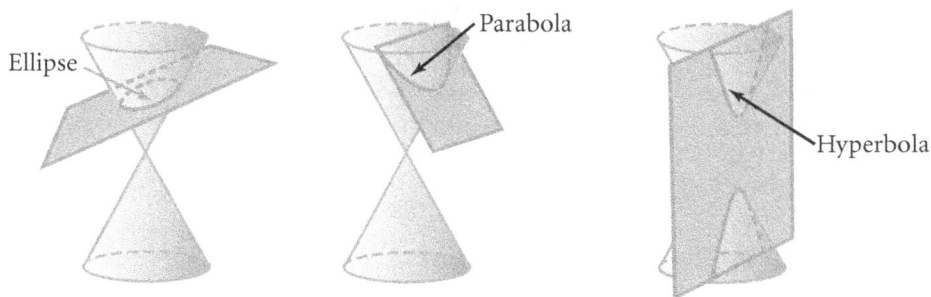

## The Parabola in the Coordinate Plane

A second geometric definition of *parabola,* which you will explore in this activity, involves the coordinate plane. This approach can be

*continued*

applied to the other conic sections as well. The definition begins by choosing a point $P$, called the **focus,** and a line $l$ (not containing $P$), called the **directrix.** The parabola defined by that point and line consists of every point $Q$ whose distance from $P$ is the same as its distance from $l$.

To illustrate this, the diagram uses the point $(0, 1)$ for $P$ and the line $y = -1$ for $l$.

The origin $(0, 0)$ is exactly one unit from $P$ and one unit from $l$, so it is on the parabola defined by this point and line.

The diagram also shows a general point $Q$, with coordinates $(x, y)$, and the segments connecting $Q$ to $P$ and to $l$. This particular point $Q$ appears to be farther from $P$ than it is from $l$, so this point $Q$ is not on the parabola.

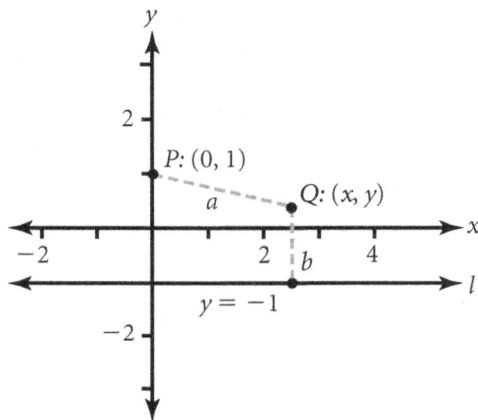

1. Estimate the distance from $Q$ to $P$ and the distance from $Q$ to $l$ to confirm that this point $Q$ is not on the parabola.

Point $Q$ is supposed to be a general point, not this specific point. Your next task is to find a condition on its coordinates that will ensure it is on the parabola. In the diagram, the distance from $Q$ to $P$ is labeled $a$ and the distance from $Q$ to $l$ is labeled $b$. Notice that segment $b$ is vertical, because this is the shortest distance from the point to the horizontal line.

2. a. Write an expression for distance $a$ using the coordinates of $P$ and $Q$.

   b. Write an expression for distance $b$ using the coordinates of $Q$ and the equation of line $l$.

3. Use your work from Question 2 to create an equation that states that the distances $a$ and $b$ are equal.

4. Simplify your equation until it looks like a familiar quadratic function.

*continued* ▸

In the example in Questions 1 to 4, point $P$ is on the $y$-axis. Also, the distance from $P$ to the $x$-axis is the same as the distance from $l$ to the $x$-axis. This makes the algebra as simple as possible.

5. Suppose the focus $P$ is the point $(2, 1)$ and the directrix $l$ is the line $y = -3$.

   a. Find a simple equation for the parabola defined by this choice of $P$ and $l$.

   b. Where is the vertex of this parabola in relation to $P$ and $l$? Why?

6. Imagine choosing some other point for $P$ and some other horizontal line for $l$. If you create and simplify the equation for the parabola defined by that choice of focus and directrix, do you think you will get a quadratic function? Justify your answer.

# Creating Parabolas

In the supplemental activity *What's a Parabola?* you were given specific combinations of a point $P$ and a line $l$ not containing that point. In each case, you had to find an equation for the parabola defined by that combination of focus and directrix.

Now you will work the other way, starting with the equation of a parabola and trying to find the parabola's focus and directrix. You may find it helpful to refer to your work on *What's a Parabola?*

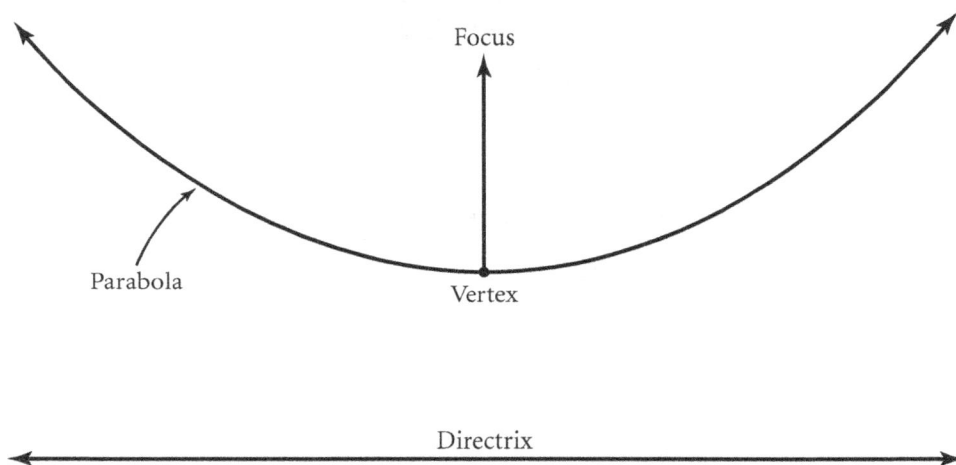

1. Find the choice of $P$ and $l$ that leads to the equation $y = x^2$.

2. Find the choice of $P$ and $l$ that leads to the equation $y = (x + 1)^2 - 2$.

3. Find the choice of $P$ and $l$ that leads to the equation $y = 4(x - 3)^2 + 1$.

# Coordinate Ellipses and Hyperbolas

In the supplemental activity *What's a Parabola?* you looked at a coordinate definition of the term *parabola*. Given a particular point *P* and a line *l* that doesn't pass through *P*, the parabola they define consists of every point *Q* whose distance from the focus *P* is the same as its distance from the directrix *l*.

Replacing "the same as" by something else leads to a way of defining the other two classes of conic sections: ellipses and hyperbolas.

Suppose *k* is some fixed positive number. Let the set *S* consist of every point *Q* whose distance from *P* is exactly *k* times its distance from *l*. If $k = 1$, you already know the set *S* is a parabola.

1. Using the point $(0, 1)$ for *P* and the line $y = -1$ for *l*, as in the diagram, and using the value $\frac{1}{2}$ for *k*, sketch the set *S*. Begin by looking for points on the *y*-axis whose distance from *P* are exactly half their distance from *l*. For more general points, you might consider whether the specific location of *Q* in the diagram would belong to set *S*, and why.

2. Use the coordinates in the diagram to write an equation stating that the general point *Q* is exactly half as far from *P* as it is from *l*.

3. Simplify your equation as much as you can.

4. Describe the shape of the graph in general terms.

5. Using the same point and line, go through the steps in Questions 1 to 4 using the value 2 for *k*. You should get points below *l* as well as points above *l*.

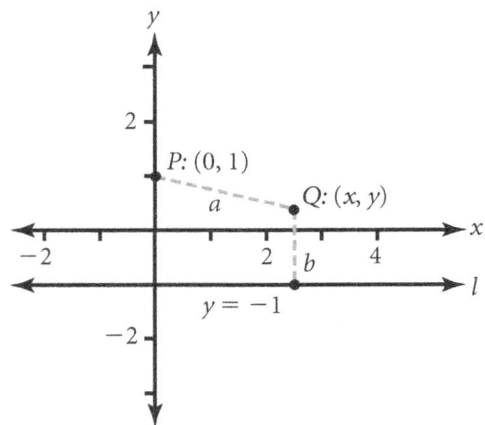

*continued* ◆

## Conic Sections Defined

The three types of conic sections—parabola, ellipse, and hyperbola—can all be defined in terms of a point $P$, a line $l$ not containing $P$, and a positive constant $k$ (called the *eccentricity*).

- If $k < 1$, the set of points whose distance from $P$ are exactly $k$ times their distance from $l$ is an ellipse.

- If $k > 1$, the set of points whose distance from $P$ are exactly $k$ times their distance from $l$ is a hyperbola.

- If $k = 1$, the set of points whose distance from $P$ are exactly $k$ times their distance from $l$ is a parabola.

In this scheme, a circle is considered a special case of an ellipse, but often the circle is described as a fourth kind of conic section.

# Another View of Ellipses and Hyperbolas

In the supplemental activity *Coordinate Ellipses and Hyperbolas*, the ellipse and hyperbola were defined starting with a point $P$ and a line $l$. There is an alternate definition based on two points.

Suppose $P_1$ and $P_2$ are two points and $d$ is a fixed positive number.

- The ellipse defined by $P_1$, $P_2$, and $d$ is the set of points $Q$ for which the sum of the distance from $Q$ to $P_1$ and the distance from $Q$ to $P_2$ is equal to $d$.

- The hyperbola defined by $P_1$, $P_2$, and $d$ is the set of points $Q$ for which the difference between the distance from $Q$ to $P_1$ and the distance from $Q$ to $P_2$ is equal to $d$.

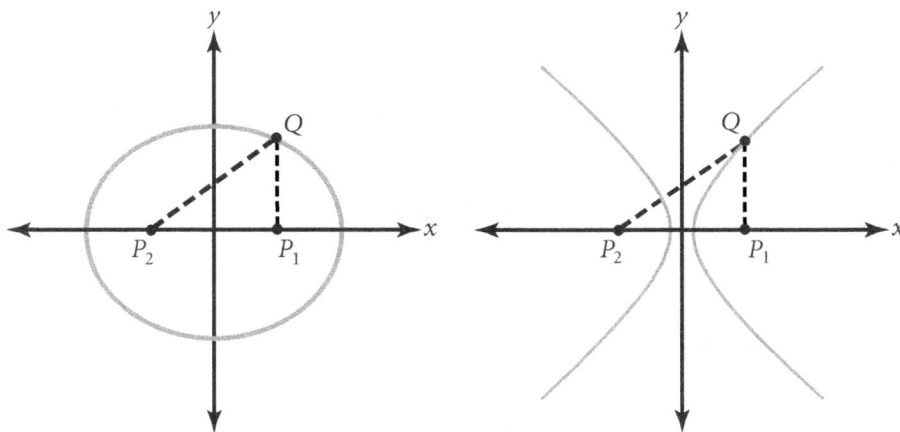

## Sketching an Ellipse

Suppose $P_1$ is the point $(2, 0)$, $P_2$ is the point $(-2, 0)$, and $d = 8$. Consider the ellipse determined by these values according to the definition just stated.

1. Verify that the point $Q = (2, 3)$ is on this ellipse. In other words, show that $P_1Q + P_2Q = 8$.

*continued* ▶

2. Find all the points on either axis that are on the ellipse.

3. Find at least three more points on the ellipse.

4. Sketch the ellipse, using your points from Questions 1 to 3 as a guideline.

## Sketching a Hyperbola

Suppose $P_1$ is the point $(2, 0)$ and $P_2$ is the point $(-2, 0)$, but this time let $d = 2$. Now consider the hyperbola determined by these values according to the definition stated earlier.

5. Verify that the point $Q = (2, 3)$ is on this hyperbola. Because $Q$ is closer to $P_1$ than to $P_2$, this means showing that $P_2Q - P_1Q = 2$.

6. Find a point on the hyperbola that is closer to $P_2$ than to $P_1$. In other words, find a point $Q$ for which $P_1Q - P_2Q = 2$.

7. Use symmetry to find two more points on the hyperbola.

8. Find all the points on either axis that are on the hyperbola.

9. Sketch the hyperbola, using your points from Questions 5 to 8 as a guideline.

It turns out that for an ellipse, $d$ must be greater than the distance from $P_1$ to $P_2$, but for a hyperbola, $d$ must be less than the distance from $P_1$ to $P_2$.

10. Explain the two parts of the last statement, as follows.

    a. For the case of the ellipse: What happens if $d$ is equal to the distance from $P_1$ to $P_2$? What happens if $d$ is less than the distance from $P_1$ to $P_2$?

    b. For the case of the hyperbola: What happens if $d$ is equal to the distance from $P_1$ to $P_2$? What happens if $d$ is more than the distance from $P_1$ to $P_2$?

# Ellipses and Hyperbolas by Points and Algebra

In the supplemental activity *Another View of Ellipses and Hyperbolas*, you saw these two definitions, based on a choice of two points $P_1$ and $P_2$ and a fixed positive number $d$.

- The ellipse defined by $P_1$, $P_2$, and $d$ is the set of points $Q$ for which the sum of the distance from $Q$ to $P_1$ and the distance from $Q$ to $P_2$ is equal to $d$.

- The hyperbola defined by $P_1$, $P_2$, and $d$ is the set of points $Q$ for which the difference between the distance from $Q$ to $P_1$ and the distance from $Q$ to $P_2$ is equal to $d$.

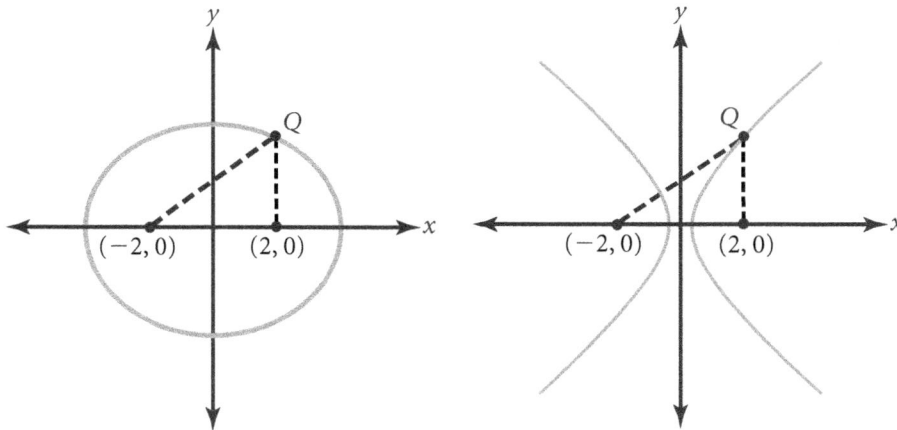

Now you will develop equations for the specific examples from *Another View of Ellipses and Hyperbolas*.

In both cases, $P_1$ is the point $(2, 0)$ and $P_2$ is the point $(-2, 0)$. Although these points make the algebra about as simple as possible, it's still very messy.

*continued* ▶

In each situation, you will likely start with an equation that involves two square-root expressions. Start by writing an equivalent equation in which these square-root expressions are on opposite sides of the equation. Then square both sides of the equation and simplify to isolate the square-root expression you still have. Then square again and simplify one more time.

1. Suppose $d = 8$. Create and simplify an equation for the ellipse defined by the points $P_1$ and $P_2$ and the constant $d$. That is, consider a general point $Q$ with coordinates $(x, y)$ and write an equation using $x$ and $y$ that says $P_1Q + P_2Q = 8$.

2. Suppose $d = 2$. Create and simplify an equation for the hyperbola defined by the points $P_1$ and $P_2$ and the constant $d$. That is, consider a general point $Q$ with coordinates $(x, y)$ and write an equation using $x$ and $y$ that says either $P_1Q - P_2Q = 2$ or $P_2Q - P_1Q = 2$.

# Generalizing the Ellipse

In the supplemental activity *Ellipses and Hyperbolas by Points and Algebra,* you saw that an ellipse can be defined as the set of all points the sum of whose distances from two fixed points is a given positive constant. Those two points are called the **foci** of the ellipse.

Suppose the ellipse is in "standard position," with its center at the origin and its foci on the *x*-axis at $(c, 0)$ and $(-c, 0)$. (The earlier activity examined the case of $c = 2$.)

The ellipse meets the *x*-axis in two points, labeled $(a, 0)$ and $(-a, 0)$ in the diagram. It meets the *y*-axis in two points, labeled $(0, b)$ and $(0, -b)$.

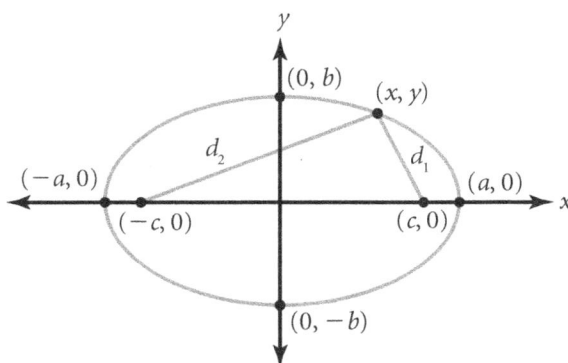

If $(x, y)$ represents a general point on the ellipse, then the distances from $(x, y)$ to the foci are given by the equations

$$d_1 = \sqrt{(x - c)^2 + y^2} \quad \text{and} \quad d_2 = \sqrt{(x + c)^2 + y^2}$$

For the special case in which $(x, y) = (a, 0)$, the sum of the distances to the foci is $2a$. Therefore the equation of the ellipse is $d_1 + d_2 = 2a$. (This shows that the case examined in *Ellipses and Hyperbolas by Points and Algebra* used $a = 4$.)

1. Show that $a^2 = b^2 + c^2$.

2. Show that the equation of the ellipse can be simplified to

$$\frac{x^2}{a^2} + \frac{y^2}{b^2} = 1$$

# Moving the Ellipse

In the supplemental activity *Generalizing the Ellipse,* you considered an ellipse whose center is at the origin and whose critical points are $(a, 0)$, $(-a, 0)$, $(0, b)$, and $(0, -b)$.

The equation for this ellipse is

$$\frac{x^2}{a^2} + \frac{y^2}{b^2} = 1$$

This ellipse is shown in the diagram.

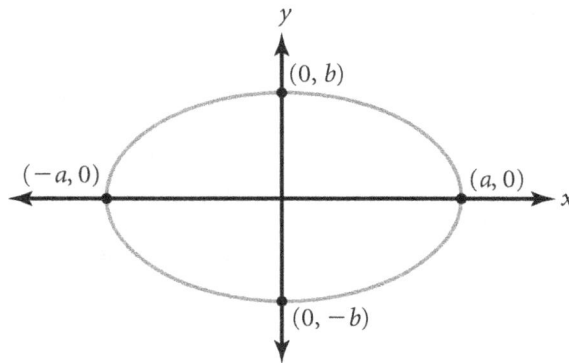

If the center of the ellipse is not at the origin, then the equation is a bit more complicated.

Suppose the ellipse with equation $\frac{x^2}{a^2} + \frac{y^2}{b^2} = 1$ has been moved $h$ units to the right and $k$ units up. Its center is now at the point $(h, k)$. If $h$ or $k$ is negative, then the ellipse has been translated left or down.

1. Find the equation of this translated ellipse. Also find the coordinates of its foci.

2. How is this equation similar to the equation of a circle with center at $(h, k)$?

# PHOTOGRAPHIC CREDITS

9 781604 400472